BLOOD OF THE ISLES

www.**books**at**transworld**.co.uk

BLOOD OF THE ISLES

Exploring the genetic roots of our tribal history

Bryan Sykes

BANTAM PRESS

LONDON · TORONTO · SYDNEY · AUCKLAND · JOHANNESBURG

TRANSWORLD PUBLISHERS
61–63 Uxbridge Road, London W5 5SA
a division of The Random House Group Ltd

RANDOM HOUSE AUSTRALIA (PTY) LTD
20 Alfred Street, Milsons Point, Sydney,
New South Wales 2061, Australia

RANDOM HOUSE NEW ZEALAND LTD
18 Poland Road, Glenfield, Auckland 10, New Zealand

RANDOM HOUSE SOUTH AFRICA (PTY) LTD
Isle of Houghton, Corner of Boundary Road & Carse O'Gowrie,
Houghton 2198, South Africa

RANDOM HOUSE PUBLISHERS INDIA PRIVATE LIMITED
301 World Trade Tower, Hotel Intercontinental Grand Complex,
Barakhamba Lane, New Delhi 110 001, India

Published 2006 by Bantam Press
a division of Transworld Publishers

A catalogue record for this book is available
from the British Library.
ISBNs 0593056523
9780593056523 (from Jan 07)
0593056531 (tpb)
9780593056530 (tpb, from Jan 07)

Maps by Red Lion Maps

Typeset in 12.5/15pt Granjon by
Falcon Oast Graphic Art Ltd.

Printed in Great Britain by
Mackays of Chatham plc, Chatham, Kent

3 5 7 9 10 8 6 4

Papers used by Transworld Publishers are natural, recyclable products
made from wood grown in sustainable forests. The manufacturing processes conform to the
environmental regulations of the country of origin.

To my son Richard, companion on very many journeys

CONTENTS

ACKNOWLEDGEMENTS

The research that led to *Blood of the Isles* was a team effort. I had a wonderful team both in the field and in the lab. Eileen Hickey, Emilce Vega, Jayne Nicholson, Catherine Irven, Zehra Mustafa, John Loughlin, Kay Chapman, Kate Smalley, Helen Chandler and Martin Richards all criss-crossed the Isles in pursuit of DNA, while Lorraine Southam, Sara Goodacre and Vincent Macaulay helped to tease out its secrets in the lab. I relied on many people's generosity in the search for our origins. The directors and staff of the Scottish Blood Transfusion Service deserve special mention for their enthusiastic backing and for their tolerance as we invaded their otherwise tranquil donor sessions. The head teachers and the staff of the very many schools we visited, particularly in Wales and Shetland, I thank for the same reasons. Talking of Shetland, I must thank Beryl Smith, who organized all our visits there in advance. But, of course, none of this would have been remotely possible without the consent and co-operation of the many thousands of volunteers who agreed to having their DNA taken and analysed.

Among professional colleagues, I am particularly grateful to Dan Bradley of Trinity College Dublin for advance access to Irish genetic data, though I should stress that I have only used published material here and also that any conclusions are my own and not necessarily Dan Bradley's. So blame me and not him. I have also benefited from the publications of Jim Wilson and Mark Thomas from University College London, who have produced very useful data from parts of Britain. Among my friends and colleagues in Oxford, William James has, as usual, been a rich source of ideas and creative conversation. I must also mention Robert Young, recently of Wadham College, who introduced me to the racial mythology of the English, a subject of which I was almost completely unaware until he sent me a reprint of his work. Norman Davies, a fellow of my own college, Wolfson, was not only a source of bountiful historical references in his magisterial *The Isles – a History* (never has a book been more thoroughly thumbed), but also helped me resolve the tricky issue of what to call my own book.

But words are not enough. Books need midwives before they see the light of day. My agent Luigi Bonomi has kept me going throughout with his irrepressible enthusiasm and I am, once again, very fortunate to have in my editors Sally Gaminara and Simon Thorogood not just consummate professionalism but great encouragement as well. Thanks too to Brenda Updegraff for her immaculate copy-editing and, as before, to Julie Sheppard who rapidly transformed my erratic handwriting into legible text.

But most of all I thank the Muse without whom nothing flows.

LIST OF ILLUSTRATIONS

First colour section
English supporter at the World Cup, 2006:
© Tony Quinn/internationalsportsimages.com/Corbis

Brutus the Trojan Sets Sail for Britain, 15th-century manuscript illumination by Master Wistace from 'The History of the Kings of Britain' by Geoffrey of Monmouth: Bibliothèque Nationale, Paris/The Bridgeman Art Library; coronation throne of Edward I: © Angelo Hornak/Corbis; Tintagel Castle, Cornwall: © English Heritage/Heritage-Images; *Merlin before King Vortigern*, manuscript illumination from 'Prophecies of Merlin' by Geoffrey of Monmouth, *c*. 1250: © The British Library/Heritage-Images; *The Last Sleep of Arthur in Avalon* (detail) by Edward Burne-Jones, 1881–98: Museo de Arte, Ponce, Puerto Rico/ The Bridgeman Art Library; Glastonbury Abbey, Somerset, detail of the ruins: © Nigel Reed/Alamy.

Head dress found at Starr Carr, Yorkshire, *c*. 7500 BC: British Museum, London; Maiden Castle, near Dorchester, Dorset, *c*. 3000 BC: Collections/Peter Thomas; inner circle of Stonehenge, Wiltshire, *c*. 2800-1500 BC: © Royalty-Free/Corbis; air view of Richborough Castle, near Sandwich, Kent, AD 43–287 : Collections/David Bowie; sword-belt buckle from the ship burial at Sutton Hoo, early 7th century, British Museum: © 2006 The British Museum; Anglo-Saxon iron helmet from the ship burial at Sutton Hoo, early 7th century, British Museum: © Visual Arts Library (London)/Alamy; detail from the Bayeux tapestry, *c*. 1080; © Nik Wheeler/Corbis.

Callanish standing stones, Lewis, Outer Hebrides, c. 3000 BC: © Adam Woolfitt/Corbis; Skara Brae, Orkney, c. 3000 BC: © Kevin Schafer/Corbis; burial chamber at Maes Howe, Orkney, c. 2750 BC: © Crown copyright reproduced by courtesy of Historic Scotland; Jarlshof, Shetland, 2400 BC: HIE/stockscotland; broch at Mousa, Shetland Islands, 1st century AD © Peter Hulme/Corbis; Hadrian's Wall, Northumberland, AD 122: © David Ball/Corbis; Pictish stone in Aberlemno, Perth & Kinross, c. AD 750: Collections/Michael Jenner.

John Beddoe, frontispiece to his book *Memories of Eighty Years*, 1910; 'A Celtic groupe', from Robert Knox *The Races of Men*, 1869; engraved portrait of Robert Knox: Wellcome Library, London; loose leaf from one of Beddoe's albums, 1882: Royal Anthropological Institute, London.

Second colour section
Adrian Targett with the skeleton of Cheddar Man, March 1997: swn.com/Darren Fletcher; drilling Cheddar man's tooth and Professor Chris Stringer with Cheddar Man jawbone: both courtesy Bryan Sykes; DNA sequence chromatogram: © Mark Harmel/Alamy.

Lindisfarne Priory and view of the seashore, Holy Island, Northumberland: both photos courtesy Bryan Sykes; Gokstad Ship, Viking Ship Museum, Oslo, c. AD 850–900: © Richard T. Nowitz/Corbis; Jarl squad, Lerwick, Shetland Isles, January 2000: © Reuters/Corbis; woman on Lewis: courtesy Bryan Sykes.

Neolithic cromlech, Carreg Sampson, Pembrokeshire: Collections/Simon McBride; Roman barracks and latrines, Caerleon, Gwent, 1st century AD: Collections/Robert Estall; Offa's Dyke near Knighton, Radnor, late 8th century AD: © Homer Sykes/Corbis; obverse of silver coin of Offa, AD 757–796: Ancient Art & Architecture Collection; aerial view of Pembroke Castle, late 12th to 13th century: © Jason Hawkes/Corbis; young Welsh fan, Canberra Stadium, Rugby World Cup, 2003: Gareth Copley/PA/Empics.

Burial chamber, Newgrange, County Meath, 3200 BC: Mike Bunn/Irish Image Collection; Mizzen Head, County Cork: Collections/Brian Shuel; Mound of the Hostages, Hill of Tara, County Meath, c. 2000 BC: George Munday/Irish Image Collection; the Broighter ship, gold, 1st century BC, found at Broighter, County Derry: Werner Forman Archive/National Museum of Ireland; Cú Chulainn: Collections/Robert Bird; spectators at the St Patrick's Day Parade, New York, 2006: © Justin Lane/epa/Corbis.

Wales Millennium Centre, Cardiff: © Photolibrary Wales; pupils at Moy School, Lahinch, County Clare: © Stephanie Maze/Corbis.

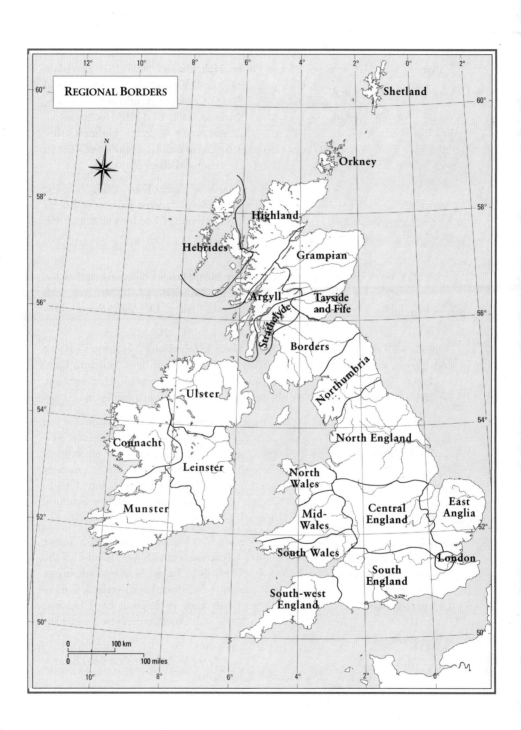

REGIONAL BORDERS

Shetland

Orkney

Highland

Hebrides

Grampian

Argyll

Tayside
and Fife

Strathclyde

Borders

Northumbria

Ulster

North England

Connacht

Leinster

North
Wales

Munster

Mid-
Wales

Central
England

East
Anglia

South Wales

London

South-west
England

South
England

0   100 km

0   100 miles

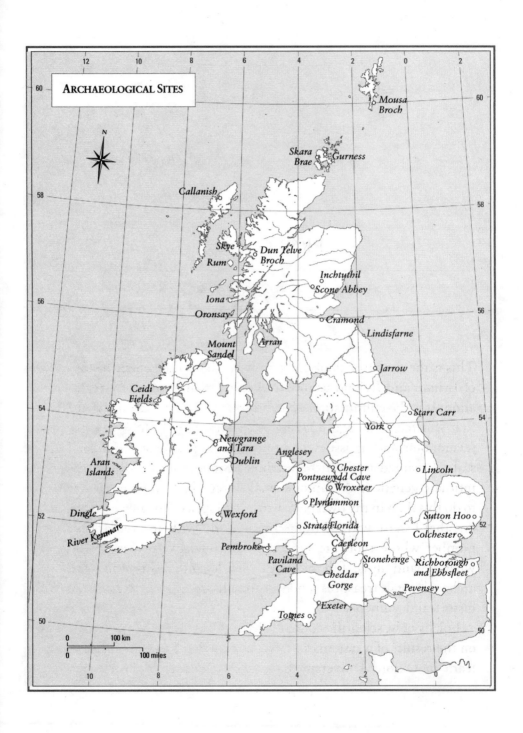

ARCHAEOLOGICAL SITES

N

Mousa
Broch

Skara
Brae

Gurness

Callanish

Skye
Dun Telve
Broch

Rum

Inchtuthil
Scone Abbey

Iona

Oronsay

Cramond

Lindisfarne

Mount
Sandel

Arran

Jarrow

Ceidi
Fields

Starr Carr

York

Newgrange
and Tara

Dublin

Anglesey

Chester
Pontnewydd Cave

Lincoln

Aran
Islands

Wroxeter

Dingle

Wexford

Plynlimmon

Sutton Hoo

River Kenmare

Strata Florida

Colchester

Pembroke

Caerleon

Paviland
Cave

Stonehenge

Richborough
and Ebbsfleet

Cheddar
Gorge

Pevensey

Totnes

Exeter

0   100 km

0   100 miles

PROLOGUE

This is the very first book to be written about the genetic history of Britain and Ireland using DNA as its main source of inform-ation. It is the culmination of an ambition, almost a dream, that I first had ten years ago. Having successfully used DNA to solve several outstanding issues about the human past on a continental scale, I wanted to push the method to its limits and dissect the intimate genetic make-up of a smaller region. And where better to do this than in my own back yard, so to speak. My own country, one that I share with 60 million others and with an even greater number whose roots are here but who now live overseas. And what a land it is, full of myth and legend, brimming with archaeological treasures and set down in a rich treasury of historical documents.

For its new, scientific content, *Blood of the Isles* relies primarily on the results of a systematic DNA survey that I and my research team in Oxford University have undertaken over the last ten years, a survey involving more than 10,000 volunteers from every part of 'the Isles'. The results, explained in the book, exceeded even my most optimistic expectations of the power of genetics to make a real contribution to our knowledge of a small region.

In *Blood of the Isles*, I approach the DNA evidence in the same way as others who write about the past using their different specialities – material artefacts, written documents, human remains and so on. The most important thing about the genetic evidence is that it is entirely independent of these other sources. It does not rely on them. However, to use genetics most effectively to fill in any picture of the past, it helps immensely to have this abundance of other evidence, and I use this resource throughout the book. Nevertheless, when you have read *Blood of the Isles* I hope you will agree that from now on genetics can take its proper place alongside history and archaeology as one of the principal lenses through which to view the past.

I have written two other books on DNA and human evolution: *The Seven Daughters of Eve* and *Adam's Curse*. You may have read them, but I certainly do not assume that you have. You don't need to in order to follow the story of *Blood of the Isles* perfectly well. However, there are some topics which are covered more extensively in the earlier books than they are here, but which to repeat here in full would be unnecessary.

I have deliberately avoided, as far as possible, putting technical data into the text. A little is absolutely essential, but too much soon disrupts the flow. For those readers who want to delve more deeply into the supporting scientific evidence I have added an Appendix (page 289) and, for real enthusiasts, I am publishing additional material on the website www.bloodoftheisles.net.

Finally, a word about the title. I use 'the Isles' rather than 'Britain' or 'UK', to avoid the pitfalls that follow from political boundaries much more recently drawn than the time depth covered by this book. Many in Ireland are not British and the Irish Republic is not part of the United Kingdom. But to leave them out would be absurd. Ours is a shared history. Throughout the book 'Ireland' includes Northern Ireland and 'Britain' embraces Scotland, Wales and England.

Some names have been changed to preserve confidentiality.

1

TWELVE THOUSAND YEARS OF SOLITUDE

Everything was ready. I selected one of the diamond-tipped bits from the sterile rack and tightened it into the jaws of the high-speed drill. Turning the dial up to 500 revolutions a second, I looked carefully to see that the spinning drill was centralized in the chuck. There must not be any mistakes, especially today. In my left hand, I picked up the jaw bone and turned it so that the outer surface of the first molar tooth was facing me. I moved the jaw under the magnifier and brought the rotating drill to within a millimetre of the enamel surface of the tooth. The tooth that had never bitten into a pizza, nor crunched a piece of celery. The tooth that I was about to drill into was 12,000 years old. The last food this tooth had touched was the flesh of a reindeer or wild horse. It was the tooth of a young man, about twenty years old when he died. This man was a hunter, one of the first people to arrive in Britain since the end of the last Ice Age.

The skeleton of the young man had been dug out of the

limestone caves of Cheddar Gorge in Somerset in 1986. Ten years later, in the autumn of 1996, I had brought his lower jaw, with the beautifully preserved teeth still embedded, to my laboratory in Oxford. I was about to attempt to recover the DNA, the genetic essence, of its original owner, trapped in the dentine beneath the hard enamel which had encased and protected it for thousands of years. As the drill made contact with the enamel surface, I steadied my left arm on the lab bench and pressed the bit into the tooth. The whining pitch of the drill came down slightly as it cut into the enamel. This was a good sign. The enamel was not too soft. That would have been a sure sign of biological decay, which would have dashed any chance of finding intact DNA. Neither was the tooth granite-hard. That would have meant that all the organic remains, including the DNA, had literally turned to stone. The Cheddar tooth was somewhere in between, neither too soft, nor too hard.

After a few seconds, the drill had cut through the enamel layer and into the dentine which lay behind. I could feel the drop in pressure as the tip of the drill moved into the softer dentine, and heard its pitch rise as the speed increased. A second or two later, I caught the scent of burning – the same unforgettable smell that instantly recalled dread visits to the dentist and the fillings of a sweet-toothed youth. It was the smell of burning teeth. This was the unmistakable scent of vaporizing protein, and the moment I caught the smell of it coming from the ancient tooth my spirits rose. From that moment on, I was sure I would find his DNA, for if the protein which was being vaporized by the drill had survived for 12,000 years, then there was every chance that his DNA would have done so too. Both are biological molecules subject to the same laws of age and decay.

As soon as I smelled the burning, I pulled across the suction line. This was a device rather like a miniature vacuum cleaner which I had rigged up for collecting the powdered dentine into a sterile test tube. With this in place, I began to drill out the dentine, carefully moving the bit up and down inside the tooth, pulling back as soon as I felt it touch the hard enamel on the other side. All the time the vacuum line was transferring the creamy white powder into the test tube and collecting it in a small pile at the bottom. Within a few minutes, I had completely excavated the inside of the tooth. In the test tube lay precisely 208 milligrams of dentine powder from the Old Stone Age.

Within two weeks, and in ways that we will cover later, I had recovered enough DNA from the Cheddar tooth to read the genetic fingerprint of its original owner – a young man whose pattern of life was so utterly different from our own that it is hard to imagine any possible connection between him and ourselves. And yet the fragment of his DNA that I had recovered from his tooth is exactly the same in every detail as that of thousands of people living in the Isles today. His descendants are with us still – and you may well be one of them.

It is now almost ten years since the day I drilled into the Cheddar tooth, but the moment is still vivid in my memory. It was not the first time I had attempted to recover DNA from ancient skeletons, but it was the most scary. This was a priceless and irreplaceable specimen. But what was I, a trained geneticist, doing drilling into the tooth in the first place? I had spent the early part of my career researching the causes of inherited diseases, mainly those affecting the skeleton – hence the location of my laboratory in Oxford's Institute of Molecular Medicine. This research had led to the discovery of the genes involved in giving strength to bones – the genes which coded for bone collagen – and to the mutations in the collagen genes which caused these often devastating diseases.

It was only a chance introduction to an archaeologist, Robert Hedges, who runs the carbon-dating lab in Oxford, that got me involved in the human past at all. Robert wanted to see if he could get more from the bone samples coming to his laboratory for carbon-dating than merely finding out how old they were. Carbon-dating relies on counting the tiny number of radioactive carbon atoms that lie in the collagen of ancient organic remains. As these atoms decay with time, the fewer there are, the older the sample. Robert got in touch, having heard about my research on the genetics of bone collagen, and we started to plan what we might be able to do with these old bones. To cut a long story short, within two years we had worked out a way of recovering DNA from human and animal bones that were hundreds or even thousands of years old.

Being the first laboratory in the world to do this, we were well placed to receive exciting samples from all over the world. Over the years we have had bits of Neanderthals; Oetzi, the famous Iceman from the Alps; various claimants to being Anastasia, the last of the Romanovs; a selection of dead poets and statesmen; not to mention the odd piece of Yeti skin. To put the DNA results from this eclectic collection into some form of context, I began a programme of collecting DNA samples from living people. For instance, although it was wonderful to be able to get DNA from the 5,000-year-old Iceman, and that became a story in itself, it only became really interesting when his DNA could be compared, and indeed matched, with someone living today. The whereabouts of his modern descendants told us something about the movement of people throughout Europe during the five millennia since his death.

Sometimes the DNA from modern people can solve long-standing riddles that had proved to be intractable by any other means. The outstanding example of this was the research on the origin of the Polynesians. These are the people who live on the far-flung islands of the Pacific. All the islands, from Hawaii in the north to Easter Island in the east and New Zealand in the far south, had been settled by Polynesians well before the time Europeans began to explore the Pacific Ocean in the early part of the sixteenth century. But where had the Polynesians come from? Was it from Asia, as the bulk of the evidence from language, domestic animals and crops suggested? Or had they arrived in the other direction from America, as the legendary Norwegian anthropologist Thor Heyerdahl believed? Like many schoolboys, I had been captivated by Heyerdahl's adventures on the balsa raft *Kon-Tiki*, on which he drifted from Peru to the Tuamotu islands, not far from Tahiti, to prove his point. So it was with a tinge of regret that, in 1995, I published the genetic data which proved conclusively that Heyerdahl was wrong. The Polynesians had come from Asia, not America. This slight regret at having dis-proved a boyhood hero was more than compensated by the proof that the Polynesians must have explored the Pacific intentionally, driving their canoes into wind and current eastwards across the vast ocean, rather than lazily drifting with the prevailing elements from South America. The ancestors of today's Polynesians were without doubt the greatest maritime explorers the world has ever known.

The proof of their true origins came from the DNA of modern Polynesians that I had collected from dozens of Pacific islands. From the detailed genetic fingerprints of the islanders I was able to trace the route that their resolute ancestors had taken through the island chains of South-east Asia and out into the vast Pacific Ocean. In ways that I will explain later, I could follow the genetic threads that had percolated through the generations and reconstruct the 3,000-year-old journeys of these astonishing navigators.

It was because I was attempting to reproduce this first success in the much more difficult arena of Europe that I found myself drilling into the Cheddar tooth. My colleagues and I had followed the same procedure that had yielded such compelling results in Polynesia. We had collected almost 1,000 DNA samples from all over Europe and, again in ways I will later explain, come to a conclusion about the origin of modern Europeans. That conclusion was, in a nutshell, that the ancestors of most native Europeans were hunter-gatherers and not, as was commonly believed at the time, farmers who had spread into Europe from the Middle East about 8,500 years ago. To say that our conclusion caused a stir is an understatement. There followed several years of fierce debate between ourselves and the proponents of the agricultural-ancestry theory, and the experiment with the Cheddar tooth was one of our efforts to prove our case. The idea behind it was that, if we could show that a very old human fossil, a genuine hunter-gatherer who lived well before farming arrived, had pretty much the same DNA as people living today, that would strengthen our side of the argument.

The fact that the Cheddar tooth DNA was identical to modern Europeans' had several ramifications. This was the DNA of a man who, without any doubt, was a hunter-gatherer who had lived at least 6,000 years before farming reached the Isles. Taken with all the other genetic evidence, the result helped to swing opinion towards a predominantly hunter-gatherer ancestry for Europeans and away from the prevalent theory of a great wave of ancient farmers sweeping out from the Middle East and overwhelming the thinly spread hunters. The heat has gone out of that particular debate by now, and I think it is fair to say that most people today think that the impact of migrating farmers on the genetic make-up of Europe was far less than previously thought.

A few months after finding the DNA from the 12,000-year-old Cheddar tooth I got permission to repeat the process with a younger specimen from the same cave. This was the famous 'Cheddar Man'. His remains had been excavated in 1903 and, like the other skeleton, had been stored in the Natural History Museum in South Kensington. They had been carbon-dated to about 9,000 years ago, still well before the arrival of farming in Britain and so still relevant to the hunter/farmer debate. Sure enough, after drilling out the tooth and analysing the DNA from the dentine powder, I could see that Cheddar Man's DNA was also thoroughly modern. It was not the same, in detail, as the earlier Cheddar tooth, but it did match quite a few modern Britons', one of whom lived just down the road from the Caves. A local television company had got wind of our work on the Cheddar fossils and, between us, we had dreamed up a format whereby, in parallel to the work on Cheddar Man's teeth, we would also test the DNA of the pupils at the local school. If we could find a DNA match between Cheddar Man and a modern-day nearby resident it would be a good local-interest story as well as a neat demonstration of genetic continuity.

With all the DNA results in from the school, and from Cheddar Man himself, the producer arranged a notorious 'reveal' session. The pupils, all aged between sixteen and eighteen, and the master who had organized the event at the school, gathered in the hall, nervously waiting for the results to be announced. The camera passed across the faces of the teenagers, each one apprehensive that it might be their DNA that had been matched to Cheddar Man. The presenter spoke, the match was revealed and the cameras swivelled round to bring one face into tight close-up.

It was not one of the pupils at all, but the history teacher who had made the arrangements – Mr Adrian Targett. Gasps all round, a blushing teacher and a score of ever so slightly disappointed teenagers.

The following day Adrian Targett's smiling face was on the front page of every national newspaper. He was pictured crouching next to the replica of Cheddar Man's skeleton at the spot in the cave where it had been discovered in 1903. Even the tabloids carried the story, impressively assembling a topless model in a skimpy rabbit-skin loincloth and with a hastily assembled flint axe. Adrian told me later that he had been offered a 'five-figure sum' to appear in a loincloth but had, sensibly, declined. The following day the story was picked up by newspapers abroad. It proved to be particularly popular in the US, probably because it fitted in nicely with the image of a bucolic English countryside in which it takes 9,000 years for someone's descendants to move 300 yards down the road. People still remember the story even now, and when I was lecturing in California last year I was introduced by the organizers as the man who got DNA from the Cheese Man.

The Cheddar Men, though they lived a very long time ago, were not the first human inhabitants of the Isles. There are scattered shreds of evidence that the Isles were once occupied by archaic species of humans, not directly ancestral to our own species, *Homo sapiens*. A shin bone from Boxgrove Quarry near Chichester on the Sussex coast, a tooth from Pontnewydd Cave in north Wales, both over a quarter of a million years old and both the remains, as far as can be told, of much sturdier, large-boned humans, more like Neanderthals than our own species. The recent discovery of flint tools that have been exposed in a crumbling cliff near Lowestoft on the Suffolk coast is evidence, albeit indirect, of a human presence on the Isles more than half a million years ago. Fascinating though these finds are, they are merely glimpses into the world of long-extinct humans who came and went but left no lasting impression on the Isles, small bands of roving hunters whose luck finally ran out. These were not our ancestors.

The earliest evidence of our own species, *Homo sapiens*, in the Isles comes from Paviland Cave just above the rocky shoreline of the Gower Peninsula to the west of Swansea in South Wales. In 1823 the Oxford palaeontologist William Buckland excavated the partial skeleton of a man. Misled by the presence of ivory ornaments near the body, Buckland assumed that he had found the remains of a woman and, because the bones were stained with red ochre as part of an unknown burial ritual, she soon became known as 'the Red Lady of Paviland'. However, a more thorough analysis of the bones, particularly the pelvis, showed that the Red Lady was actually a man, though he still retains the title. When Buckland found these bones they were so well preserved that he thought they could not be all that old. His theory was that they were the remains of a woman who had been living in the cave while working at a nearby Roman camp. But he was wrong again. We now know from carbon-dating that the Red Lady was much older than the time of the Roman occupation. 'She' died 26,000 years ago and 'her' pendant was not made of elephant ivory but had been carved from the tusk of a mammoth. We know, from the deliberate burial, that the Red Lady was survived by her relatives, but no trace of them remains. After the time of the Red Lady, there is a long empty gap in the fossil record of the Isles. There is nothing until the time of the 'older' of the Cheddar Men, just over 12,000 years ago. Why the break? There is one very simple answer – the Ice Age.

About 24,000 years ago the temperatures in the northern latitudes around the globe, including the Isles, began to drop as the planet entered once again into the downward phase of a glacial cycle. These regular cycles of bitter cold and comparative mildness have been going on for at least 2 million years. They are caused by the slight shifts in the way the earth rotates and moves in its orbit around the sun. The shape of the orbit changes from circular to elliptical and then back to circular about once every 96,000 years. The angle of the earth's axis changes, shifting the positions of the Arctic Circle and the Tropic of Cancer up and down by 3 degrees of latitude, and several hundred miles, once every 42,000 years. Another cycle, every 20,000 years, alters the seasons when the earth is at different parts of its orbit. As the earth runs through this cycle, the signs of the zodiac slowly move round and we enter new astrological 'ages', the latest being Aquarius. The combination of all three cycles one on top of the other means the earth's climate never stands still for long. The effect is to change the amount of sunlight which hits the higher latitudes in both hemispheres, slowly increasing and decreasing as the overlapping cycles gradually shift the planet's position with respect to the sun. We are now in a warm phase of the long-term glacial cycle, but it will not last for ever and at some as yet unpredictable time in the future we will slide inexorably into another Ice Age. How soon the next cold phase will begin and to what extent its chilling effects will be tempered by 'global warming' are all uncertainties for future generations.

For the descendants of the Red Lady and the other scattered occupants of the Isles 24,000 years ago, even though they did not know it, their tenancy of the land was coming to an end. Gradually the year-on-year temperatures began to fall. Snow that covered the mountains in winter no longer melted in the summer and gradually built up into a permanent ice cap. The sea began to recede as more and more water became locked in permanent ice sheets, not just in the Isles but also at the Poles and over the mountain ranges of Europe, Asia and America. The Isles became a peninsula as the North Sea receded. Britain and Ireland were joined. Vicious winds howled around the edges of the expanding ice cap as the weather systems shifted away from the succession of moisture-bearing Atlantic depressions towards an Arctic climate of intense, dry cold. And all the time, the ice moved south. The herds of migrating game – reindeer, bison, wild horse and mammoth – moved their ranges away from the worsening conditions, and the scattered groups of humans who depended on them for food had no choice but to follow them. By the time of the coldest phase of the Ice Age, 18,000 years ago, there were no humans left in Britain, or anywhere else in Europe north of the Alps.

The descendants of the Red Lady and their contemporaries had retreated to refuges in southern France, Italy and Spain, abandoning northern Europe to the frost and ice. Great glaciers flowed downhill from the ice domes over the mountains of northern Britain, gouging out steep-sided valleys and pulverizing the bedrock as they ground their way across the landscape, obliterating everything in their path. All evidence of human occupation in northern Britain was completely erased by the ice. Only south of a line from the English Midlands to central Ireland, which marked the edge of the ice, could any trace remain.

And then, quite suddenly, the climate began to improve as the planet moved its alignment in the heavens. The warmth of the sun returned to the northern latitudes and the ice began to melt. Our ancestors followed the herds north from their huddled refuges as the frozen land began to thaw. Carbon-dating of charcoal left by campfires has traced the advancing front and by 13,000 years ago they had reached northern France. A millennium later, the older of the Cheddar Men, or his immediate ancestors, were among the first to arrive in the Isles, by foot across the land that now lies beneath the North Sea. His are among the oldest remains to be found anywhere in post-Ice Age Britain. He arrived in a landscape scrubbed clean of human occupation by the effects of the Ice Age, even though the ice itself never reached as far south as his home in Cheddar. His camp in the gorge was perfect as an ambush site to trap the migrating herds of reindeer as they moved from their summer feeding grounds on the high Mendips to spend the winter on the Somerset Levels. Remains at the site showed he was skilled at making the variety of flint tools on which the life of the hunter depended.

When he arrived, 12,000 years or so ago, the Isles were connected to each other and to the rest of continental Europe. The sea was 100 feet lower than it is now and large tracts of land that are now under water were well above sea level. Ireland was connected to mainland Britain through a broad plain that joined it to the west coast of Scotland and took in what is now the whisky isle of Islay. The Irish Sea, which now entirely separates Ireland from the rest of the Isles, was then a narrow sea inlet between flat plains, blocked at its northern end by the isthmus that joined Scotland to the north of Ireland. The Western Isles off the north-west coast of Scotland were similarly joined to the mainland with a narrow strip of dry land. The Hebridean islands of Skye, Mull, Rum, Coll and Tiree were not islands then; neither were the Orkney Islands, now separated from the far north of the Scottish mainland by the turbulent seas of the Pentland Firth. Only the Shetland Isles, 60 miles north of Orkney, were truly islands in those far-off days.

Most important of all, there was dry land connecting Britain to continental Europe. This was no narrow causeway, but a wide rolling plain joining eastern Britain to the rest of Europe from the Tyne in the north to Beachy Head near Eastbourne in the south. The entire southern section of what is now the North Sea was dry land intersected by wide rivers. The Thames was then a tributary of the Rhine, their joined waters emptying into the sea 100 miles east of Newcastle upon Tyne. What is now Britain and Ireland, separated by shallow seas, was then a great peninsula protruding into the Atlantic Ocean. The Irish Sea was open only at its southern end and the North Sea was dry land. The sea level was rising as the global temperature climbed back up after the last Ice Age. The great ice sheets that covered the northern hemisphere were melting, as their remnants in the polar north continue to do today.

Even now, the Isles are still twisting and turning in the aftermath of the last Ice Age. The immense weight of snow and ice across the Scottish Highlands and, to a greater extent still, over the mountains of Sweden and Norway pushed these lands further down into the semi-molten layers deep beneath the earth's solid crust. The continents are floating on molten magma and, just like a cargo ship, they move up and down depending on their loadings. Unlike a ship, when one part goes down under the weight of ice, a nearby section can be pushed up. This happened to the land that now lies beneath the North Sea and which was artificially elevated by the sinking of Scandinavia under the weight of its ice dome. As this melted, the sea level certainly rose, simply because there was much more water in the oceans. But the land that had been pushed down into the magma began to rise up again as the load of ice melted away. And as Norway and Sweden floated back up towards their pre-Ice Age levels, so Doggerland, as this now vanished land beneath the North Sea is called, sank back down, quickening its submergence beneath the rising sea and accelerating the conversion of the Isles from peninsula to the real thing.

The Isles are still slowly convulsing. The southern and eastern coasts of England are sinking at the rate of an inch a year, causing the kind of coastal erosion that has shed the flint tools half a million years old from the crumbling cliffs of Suffolk. The west and north coasts are rising at about the same rate or sometimes faster. Dunbar, on the east coast of Scotland not far from Edinburgh, for example, is rising at the rate of 2 inches a year. Ireland is still lifting itself from the ocean.

By the time the 'younger' Cheddar Man lived in the same gorge, 9,000 years ago, the climate had improved even more, but the great herds of big game had gone. What had once been open tundra was now dense woodland and, though the climate was warmer, the business of living was very much harder. His diet was opportunistic: fish and crayfish from the river that tumbled out of the rocks; birds and mammals, perhaps a squirrel or a pine marten from the woods; and, on a good day, a red deer. In the autumn there were mushrooms and nuts from the woods around the gorge. Woodland was now the dominant feature of the landscape of the Isles as the chilling effect of the Ice Age wore off. Though the types of trees varied from place to place, the land was covered in dense forests from the southern shores to the glens of the northern Highlands. Only the uplands remained free of natural tree cover due to a combination of altitude and poor soil.

Since the time of the 'older' Cheddar Man, the Isles have been almost continuously occupied. During the millennia that his remains slept in the silent limestone caves of Somerset, almost everything has changed in the Isles. The landscape has been transformed from open tundra to thick forest to cultivated fields.

Where once he hunted for food, tourists throng the gorge and queue for cream teas. From a total population of a few thousand in the Stone Age, the Isles are now home to more than 60 million people. Beyond the shores, a further 150 million people from all over the world can trace their roots back to the Isles. While his bones were gradually entombed by the drip, drip, drip of lime-stone water in the silence of his cave, the ancestors of the ancient Celts have arrived in Wales and Ireland, the ground has trembled under the marching feet of Roman legions, the shingle beaches of Kent have yielded to the keels of Saxon warships, and the blood-curdling cries of Viking raiders have echoed from the defenceless monasteries of Northumbria and the Scottish islands. While he endured 12,000 years of solitude, the world outside pulsed with life – and death. His DNA stayed where it was, but outside the cave it had another life in the generations of descendants whose stories we can now begin to unfold.

2

WHO DO WE THINK WE ARE?

On Easter Day 1278, Edward I, King of England, accompanied by Queen Eleanor and a glittering retinue of knights and ministers, arrived at the Benedictine monastery of Glastonbury in Somerset. The reason for his visit was very specific – and very deliberate. He and his court were there to open the tomb of the legendary King Arthur. In a lavishly elaborate ceremony, two caskets containing the bones of Arthur and his queen, Guinevere, were taken from the tomb, the bones removed and carefully laid out on the altar of the monastery chapel. The following day Edward wrapped Arthur's bones in sheets of silk and solemnly placed them in a painted casket decorated with Arthur's portrait and his coat of arms. Queen Eleanor then mounted the platform and performed the same rites with the bones of Guinevere. After this the caskets were placed in front of the high altar and the royal party departed.

What was Edward up to? Why did he go to so much trouble to

travel all the way to Glastonbury? He was there for one very simple reason. He was aligning himself with the legend of King Arthur and through him laying claim to the ancient kingdom of the Britons. He was able to capitalize on the predominant myth about the origins of the British people, a myth that utterly dominated the Middle Ages. We may believe that nowadays we are beyond the grasp of hazy origin myths and treat them as the sole preserve of ignorant and primitive people clinging to absurd notions of their past. But in my research around the world I have more than once found that oral myths are closer to the genetic conclusions than the often ambiguous scientific evidence of archaeology. Hawaiki, the legendary homeland of the Polynesians, was said to be located among the islands of Indonesia, and genetics proved it. The Hazara tribe of north-west Pakistan had a strong oral myth of descent from the first Mongol emperor, Genghis Khan, and his genes are still there to this day. These are just two examples.

Only when I began my research in the Isles did I come to appreciate that we are just as entangled in our own origin myths as everybody else. They are still very powerful and, as in other parts of the world, they may contain grains of truth that we can test by genetics. I believe we are just as vulnerable to the power of myth about our own origins as the Polynesians or the Hazara or, indeed, the witnesses to the elaborate ceremony at Glastonbury over 700 years ago. The modern historian Norman Davies castigates archaeologists for their over-materialist approach to the past and their disdain for myth. I am on his side. While no one would be foolish enough to suggest that they are entirely accurate in every detail, myths have a very long memory. They are also extremely influential. To see how our own origin myths have developed, let us return to the Middle Ages.

The legend of King Arthur was brilliantly exploited by Edward I and many other of the Plantagenet kings who reigned during the Middle Ages. By linking himself to the mythical dynasty of ancient British kings he was seeking to justify his attempts to become sovereign of the whole of Britain. Twenty years after Glastonbury, he argued in the papal court in Rome that his descent from Arthur and a long line of ancient British kings gave the English crown rights over Scotland and had been ample justification for his military campaigns. His great-grandfather, Henry II, had done a similar thing when he arranged for Arthur's remains at Glastonbury to be 'discovered'. His grandson Edward III showed his enthusiasm for the myth in 1348 by instigating his own version of the Arthurian Knights of the Round Table: the Order of the Garter, a select company of twenty-four grandees that still continues today. It is no co-incidence that Prince Charles, the current heir to the throne, was christened Arthur among his many names.

The origin of the myth itself was a quite brilliant work of creative imagination by a Welsh cleric, Geoffrey of Monmouth, writing in 1138. *The History of the Kings of Britain* has everything an origin myth should have. It is full of heroic deeds, terrible battles, black treachery, and is woven with just enough threads of authenticity to be taken seriously. It even had its own mysterious source – a book (never discovered) 'written in the British language which told of the lives of the ancient British Kings from Brutus, the first, to Cadwallader, the last', given to Geoffrey by a mysterious archdeacon, named Walter, in Oxford where he wrote the *History*.

It is hard for us, in retrospect and living in a world where most mythologies, or so we like to think, require at least a semblance of supporting evidence, to believe that Geoffrey's *History* should have been taken quite so literally. But what gave the *History* such an enduring influence, apart from its use for political advantage, was that woven into the improbable narrative and sheer fantasy were crumbs of credible historical fact. It was very specific about the tide of events and enjoyed huge popularity because of its enthusiastic endorsement by a succession of royal dynasties. It became, quite literally, a medieval bestseller, and as its popularity increased, so the myth it created slowly transubstantiated into objective truth. It was believed every bit as much as the Greeks were certain of their Olympian pantheon of Zeus, Apollo, Athena and Poseidon.

Geoffrey begins his *History* with a description:

Britain, best of Islands, formerly called Albion is situated in the Western Ocean, between Gaul and Ireland. It is in length 800 miles, in breadth 200 and is inexhaustible in every production necessary to the use of man. For it has mines of all kinds, the plains are numerous and extensive, the hills high and bold and the soil well adapted to tillage, yields its fruits of every species in their season. The woods abound with a variety of animals and afford pasturage for cattle, and flowers of many lines, from which the eager bees collect their honey. At the bases of their mountains that tower to the skies are green meads, delightfully situated, through which the pure streams flow from their fountains in gentle soothing murmurs. Fish also live in abundance in the lakes and rivers and in the surrounding sea. It is inhabited by five differ-ent nations, Britons, Saxons, Romans, Picts and Celts. Of these the Britons formerly, and prior to the rest, possessed the country from sea to sea until divine vengeance because of their pride, they gave place to the Pictish and Saxon invaders. In what manner and whence they came will more fully appear in what follows.

According to the *History*, the very first inhabitants of Britain were a race of giants under Albion, a son of the sea-god Poseidon. Albion and the other giants were the children of a band of fifty women who arrived in the empty land having been banished for killing their husbands. There being no men, the fifty women mated with demons to conceive their giant offspring. The demise of Albion came about when he joined forces with two of his brothers to steal, from Hercules, the herd of cattle he had been sent to capture in Spain as the tenth of his twelve labours. Albion and his giants ambushed Hercules as he was passing through the south of France on his way home to Greece with the cattle. Hercules fought off Albion, aided by his father Zeus who arranged for a shower of rocks to fall from the sky at just the right moment, and slew the giants. After that defeat, though the giants continued to inhabit Britain for the next 600 years, their numbers dwindled until only a few remained.

Already this is a rich history, firmly linked for the benefit of the readership to the classical mythology of Zeus, Poseidon and Hercules. The next arrivals were no less well connected to the classical world and came to Britain as a direct result of the Trojan War. When Troy fell to the Greeks, Aeneas and a group of his followers escaped and made their way to Italy, where they established the settlement that was to become Rome. The link between Troy and Britain begins with the birth of Aeneas's grandson, Brutus. The soothsayers, indispensible contributors to all good mythologies, predict that he will cause the death of his parents. Which, of course, comes to pass. His mother dies in childbirth and he accidentally shoots his father. A deer runs between the young Brutus and his father while they are out hunting. Brutus fires the arrow, which glances off the deer's back and hits his father in the chest. After this misfortune Brutus is banished. His wanderings take him to Greece, where he precipitates a revolt by slaves descended from Trojan prisoners of war, and liberates them. Looking for a new home, they sail to a small deserted island, where Brutus finds a temple dedicated to the goddess Diana. In a dream Diana reveals to Brutus the existence of a great island past the Pillars of Hercules (the Strait of Gibraltar) and out into the ocean towards the setting sun.

> Brutus, there lies in the west, beyond the realms of Gaul, an island surrounded by the waters of the ocean, once inhabited by giants, but now deserted. Thither go thou, for it is fated to be a second Troy to thee and thy posterity; and from thee shall Kings descend who shall subdue the whole world to their power.

Though the island is inhabited by giants, Diana reassures Brutus that, following their defeat by Hercules, they are few in number and easily overcome. Once there, Diana promises him, Brutus will build a new Troy and found a dynasty of kings that will eventually become the most powerful on earth. You can already see how Geoffrey has cleverly sculpted his *History* to make it irresistible for any British king to claim this mantle for himself.

Now on a divine mission, Brutus sets sail for Albion with his Trojans. All ancestors, whether mythical or entirely real, must place their first foot on dry land somewhere. Brutus chose Totnes in Devon, a few miles up the River Dart from the open sea. The rock on which his foot first made contact with Albion is still there. Brutus and his men made short work of the giants and set about exploring the virgin country. Their chosen site for New Troy was on the River Thames. New Troy, or Troia Nova, became Trinovantum and, later, London. Another stone, still visible today in Cannon Street near the City's financial quarter, was the altar that Brutus built to honour Diana whose divine guidance led him to Albion. Thus it was, according to the *History*, that Brutus, grandson of Aeneas of Troy, became the first king of Britain.

Twenty years after he first stepped ashore at Totnes, Brutus died and Britain was divided into three parts, England, Scotland and Wales, each ruled by one of his three sons in that order of seniority. When the two younger sons died, the whole island reverted to the eldest, Locrinus. It was his alleged direct descent from Locrinus that Edward I used as the justification for his military campaigns against both Wales and Scotland in the late 1200s. For Edward, it was entirely legitimate to restore the whole of Britain under one crown – his, of course.

From Brutus and Locrinus, a long line of kings trickles down through the centuries, a rich vein of quasi-historical material for mythologists and authors. Shakespeare's inspiration for King Lear came from this list. Another legendary king was Lud, who rebuilt the walls of New Troy; it was through corruptions of Lud's name that it eventually became London. After Lud, the next in line was Cassivelaunus, whom we shall meet again later on. It was during his reign that Julius Caesar launched his military expeditions in Britain in 55 and 54 BC. And these were certainly not mythical. Caesar was well aware of the legend of common descent of both Romans and Britons from Aeneas and the Trojans. But this did not alter his view that the Britons, during their long centuries of isolation, had become degenerate and lost their skill in the art of war.

The full-scale Roman invasion launched by Claudius in AD 43 reduced the power of the British kings but did not extinguish it. But it is the events in the centuries after the Roman occupation ended and what the myth has to say about the Saxons that give it its greatest modern significance. Not all British kings were heroes in the *History*, and as Roman power in Britain declined in the early fifth century AD, the country became the focus of Anglo-Saxon ambitions. At this point the crown passed to the ambitious and treacherous tribal chieftain Vortigern.

After the death of the rightful king, Constantine, Vortigern arranges for the coronation of Constantine's unworldly son, Constans, in exchange for his own effective management of the country. But that is not enough for Vortigern, and he orders Constans's murder. Even then he is not crowned, but assumes the title of King of the Britons. Constans's brothers, the rightful heirs, flee to Brittany and prepare for an attack to regain the crown. To protect himself against the forthcoming war, Vortigern makes the fateful decision to recruit outside help. According to the *History*, he sights three ships in the Channel which, he discovers, are manned by Saxons under their leader Hengist. They have been sent to seek settlements of their own, their homeland being no longer able to support them – an exercise carried out, apparently, once every seven years.

Vortigern promises the Saxons land on which to settle in return for their military support in the anticipated war and they receive, first, part of Lincolnshire, then, after Hengist gives his daughter Rowena to the infatuated Vortigern, he receives the earldom of Kent. Appalled at Vortigern's gift of land to the Saxons – none more so than the dispossessed Earl of Kent – the Britons make Vortimer, Vortigern's son from his first wife, their king and expel the Saxons from the shores. But Vortimer is poisoned on Rowena's orders and Vortigern is made king once more. The Saxons return in force. Hengist convenes a great assembly of British earls and barons under Vortigern's patronage to thrash out the peaceful integration of his Saxons in Britain. In the spirit of the meeting, everyone arrives unarmed. But the treacherous Hengist orders each of his men to conceal a long knife in their clothes. On a prearranged signal, each Saxon pulls out his knife and kills the Briton standing next to him. Only one man survives to tell the tale. The Saxons banish Vortigern, no longer useful to them, to Wales and take possession of England.

The mythology surrounding the arrival of the Saxons was completely transformed in later centuries, but for Geoffrey of Monmouth it began through an act of treachery and betrayal. It is against this background that the greatest hero of the *History*, Arthur, makes his appearance. The wretched Vortigern retreats to the Welsh hills, but his attempts to build himself a fortress are frustrated by the collapse of each day's work during the following night. He is told by his court bards that only by mixing the blood of a child with no father into the mortar will this nightly collapse be avoided. His men are despatched to all parts of Wales to discover such a boy; in Carmarthen they find one and bring him, with his mother, to Vortigern. But this is no ordinary boy: it is Merlin. He challenges Vortigern's bards as to why they think it necessary to sacrifice him to build the castle. What is it, he demands, that lies beneath the site to make it unstable? They cannot answer. Using his own magic powers to see into the ground, he tells Vortigern that if he excavates the soil beneath the castle site he will discover a subterranean pool. Vortigern's men dig down and, sure enough, there is the lake. Drain the pool, Merlin prophesies, and you will find two hollow stones, each containing a sleeping dragon. The pool is drained and the dragons, one red, one white, awake and begin to fight. At first the White Dragon prevails but is eventually overcome by the Red Dragon. The White Dragon symbolizes the Saxons, the Red Dragon the Britons. The message is clear. Fight back against the treacherous

Saxons and you will prevail. Even today the Red Dragon, and all it stands for, is prominent on the Welsh flag and other national emblems – a direct legacy from Geoffrey of Monmouth.

The ultimate victory of the Britons over the Saxon invaders is a recurring theme throughout the *History* and no character symbolizes this resistance more than King Arthur. But his birth is not without its own dark side. If Vortigern's infatuation with Rowena sowed the seeds of his downfall and the invasion of the Saxons, it was infatuation that led to Arthur's birth.

Merlin, who is by now living by his uncannily accurate prophecies, foretells that Uther, the younger of Constantine's two surviving sons, will become the next King of the Britons. Returning from exile in Brittany (and landing at Totnes – always a good start), Uther and his brother beat back the Saxons, killing both Vortigern and Hengist in the process. But the elder of the two is poisoned, again as prophesied, and on his death a comet appears, at the head of which is a ball of fire resembling a dragon. Merlin, conveniently on hand, interprets this as a sign that the younger brother must be crowned king. Thus Uther becomes Uther Pendragon, Uther of the Dragon's Head, and King of the Britons at the same time.

At Uther Pendragon's coronation and victory celebration in London arrive Gorlais, Earl of Cornwall, and his wife Eigr, the most beautiful woman in Britain. That is when the infatuation begins. Tired of the attention that his wife is getting from Uther, Gorlais takes her from the palace and sets out home for Cornwall and his newly built castle at Tintagel among the high sea cliffs. Uther commands that Gorlais return to London at once and when he refuses Uther follows him to Tintagel. The only entrance to the castle is over a narrow and easily defended cause-way, along which only one man can pass at a time. Uther appeals to Merlin for help and Merlin transforms him into Eigr's husband, in which disguise he enters both the castle and her bed, where Arthur is conceived. That same night, Uther's soldiers capture and kill the real Gorlais.

In Arthur, Geoffrey's *History* has constructed the most enduring of British mythical heroes. With scant reliable historical material to go on – or, it must be said, to get in the way of a good story – Arthur's exploits are so familiar that they scarcely need repeating here. But it was the extravagance of Arthur's adventures that sowed the seeds of the *History*'s eventual demise.

Following his father's death (another poisoning), Arthur is crowned at the age of fifteen. Immediately after the coronation, he sets off on a spree of military conquest, first in Britain, then abroad. At first he defeats the few remaining Saxons, then pushes the encroaching Picts back to northern Scotland before invading in turn Ireland, Iceland, Norway, Denmark and the French territories of Normandy and Aquitaine – all in the space of nine years. At the celebrations to mark his return to Britain, Arthur receives a message from the Roman Emperor demanding his submission and the payment of tribute. Incensed by this insult, he sets off for Italy at once to demand his own tribute from Rome, taking the city in the process. While he is away, he is betrayed by his treacherous nephew, Mordred, who seizes both the crown and Arthur's queen, Guinevere. Arthur returns and kills Mordred at the battle of Camlan, in Devon, but is himself wounded – though not killed.

At this point, Geoffrey's *History* becomes strangely cloudy. While he is perfectly content to detail the death of the other ninety-eight kings in his account, when it comes to Arthur himself the ending is left deliberately vague. According to Geoffrey, Arthur is taken to the idyllic Isle of Avalon to have his wounds tended. Then comes the briefest of statements: 'This is all that is said here of Arthur's death', though the year AD 542 is noted – the only date in the entire *History*. After giving over almost a third of the book to every detail of Arthur's life, an ending so abrupt and so inconclusive may come as a surprise. But this kind of ending is nowadays very familiar, especially where there is even the remotest possibility of a sequel. Could it be that Geoffrey of Monmouth found it impossible to kill off his most important creation? Like so much about Arthur, we will never know. In fact, Geoffrey did write another book on his other famous creation, *Vita Merlini – The Life of Merlin* – and in this he does elaborate a little on Arthur's arrival at Avalon accompanied by his entourage:

After the battle of Camlan we took the wounded Arthur to Avalon. There Morgen [Morgan La Fay] placed the king on a golden bed, and with her own noble hand uncovered the wound and gazed at it long. At last she said that health could return to him if he were to stay with her for a long time and wished to make use of her healing art. Rejoicing, therefore, we committed the king to her, and returning gave our sails to the favouring winds.

As Geoffrey's book became more and more popular, the ambiguity about Arthur's uncertain fate became a problem for the Plantagenet kings who so cleverly used the *History* to link themselves to the ancient line of British kings. What if Arthur were still alive, even after 700 years? And if he were, could he return? Not long after Geoffrey's *History* appeared, King Henry II, in his campaigns to subdue the Welsh, became so concerned that the slightest possibility of Arthur's miraculous reappearance would encourage resistance that he decided to do something about it. It was Henry who arranged for the remains of Arthur and Guinevere to be 'discovered' when Glastonbury Abbey was being rebuilt after a fire. And it was Henry who, in a move to reinforce the genealogical connection to the mythical dynasty which the Plantagenets claimed, had his own grandson christened Arthur in 1187. Not only that, when the boy attained the throne he was to be known not as Arthur I but Arthur II. His uncle, King John, put an end to that when he arranged young Arthur's murder in France when he was sixteen.

A century later, Henry's great-grandson, Edward I, played the Arthurian connection for all it was worth, letting it be known during his campaigns in Wales that he was personally fulfilling Merlin's prophecy that Arthur would be reincarnated. Annoyingly for Edward, exactly the same claim was being made by his principal adversary in Wales, Llywelyn ap Gruffydd.

The legends of Arthur and Merlin have always been particularly popular in Wales. At the end of the fifteenth century, Henry Tudor, later Henry VII, used the myth very effectively in his campaign for the crown and his defeat of Richard III. To emphasize the connection, he campaigned under the banner of the Red Dragon and christened his eldest son Arthur. But alas, like Henry II's grandson of the same name, this Arthur never made it to the throne either. He died of consumption at the age of fifteen, seven months after marrying Catherine of Aragon, who later became the first of his younger brother Henry VIII's many wives.

Eventually, through repetition and royal patronage, Geoffrey's *History* became the foundation for the myth that sustained and defined racial aspirations and ambitions for half a millennium. Even now, the division between Saxon and Briton (for which also read Welsh or Celt) that is such a feature of the *History* is still not far beneath the surface. The Britons, personified by Arthur, are the truly indigenous people of the whole of Britain and the Saxons are treacherous impostors. The reign of Henry VIII saw this mythology begin to undergo at first a subtle and then a dramatic and sinister transformation.

At first, the legacy of the *History* went from strength to strength, becoming under Henry VIII a vital argument in his struggle with the Pope to divorce his first wife, Catherine of Aragon, in order to marry his second, Anne Boleyn. Henry's ambassador to the papal court, the Duke of Norfolk, used the genealogical claims to the ancient line of kings to assert Henry's supreme jurisdiction in his own realm and to back the claim that he did not need Rome's permission for anything. Norfolk told the bemused court that the *History* recorded how a British king, Brennius, had once conquered Rome, that the Roman Emperor Constantine himself was also on the list of kings, and that Arthur had been Emperor of Britain, Gaul and Germany. These arguments made little impression in Rome, which continued to resist the divorce, but they featured strongly in the laws passed by Henry to enact the break with the Roman Catholic Church and to establish the Church of England with Henry at its head. Sovereigns today still assume that title.

But even at this moment of triumph for the myth, it was being undermined. Henry VII, eager that his new dynasty should appeal to the long-established monarchies of Europe, had commissioned an Italian scholar, Polydore Vergil, to write a new history. Henry VII died before it was finished but Henry VIII allowed work to continue and it was finally completed in 1513. As a Renaissance scholar, Polydore Vergil was trained to do what few historians had done before – look for the evidence. When he came to scrutinize Geoffrey's *History*, it was soon very clear that there was hardly any. His main source, the mysterious book that Geoffrey had been given by Walter, Archdeacon of Oxford, was never found. Even worse, there was no mention of Geoffrey's principal hero, Arthur, in any other histories.

One of these, *The Ruin of Britain*, a sixth-century polemic by the Breton monk Gildas, was always thought to be extremely shaky, but another, Bede's *History of the English Church and People*, published in AD 731, is a far more serious and reliable account – and does not mention Arthur at all. Surely, argued Polydore Vergil, it is inconceivable that a serious history such as Bede's could have failed even to mention a king who had not only regained territory from the Saxons and Picts but had also conquered Ireland, Iceland, Norway, Denmark and much of France only 200 years previously. This was, it had to be admitted, a strong and rational argument. But myth and reason do not necessarily concur. So closely was the court of Henry VIII wedded to, even dependent on, the antiquity of their dynastic claims as 'authenticated' by Geoffrey of Monmouth, that the King refused to allow Vergil's work to be published for a further twenty years. That it was published at all was a sign, not of the triumph of reason, but that the myth itself was beginning to lose its political usefulness to the King. The ancient Britons and their affinity to the Catholic Church were becoming an embarrassment.

But the popularity of Geoffrey's *History* was still sufficient for the publication of Polydore Vergil's alternative *Anglica Historica* to be greeted with outrage and the author condemned as an unscrupulous papist who had set out to undermine the new self-confidence of the English Church. Even during the reign of Henry's daughter, Elizabeth I, the *History* was still a source of inspiration to poets like Edmund Spenser, whose *Faerie Queene* links Elizabeth and Arthur, and, of course, to Shakespeare, whose *King Lear* draws its characters straight from Geoffrey of Monmouth.

Nevertheless, the currency of the myth among scholars continued to decline, even though it enjoyed a brief revival of royal enthusiasm in 1601 when the Stuart King James VI of Scotland/ I of England cast himself as the embodiment of Merlin's prophecy and the restorer of the ancient unity of England, Wales and Scotland first won by Brutus and Locrinus. Eventually, the Stuarts were overthrown and the crown passed to William of Orange. Though he, one would have thought, could not possibly claim a link to the myth, coming as he did from Protestant Holland, this did not stop the poet R. D. Blackmore from portraying William as the Christian Arthur. Even more bizarrely, he managed to twist the myth to the point where William became the champion of the true religion of the ancient Britons (Protestantism!) against the heathen Saxons (Catholic!). That shameful episode was the last bow of the myth on the political stage, though its popularity even today is witness to its continuing fascination.

The real reason for the slow decline of the myth of a united, essentially Celtic Britain with ancient foundations, as elaborated in the *History*, was that, following the Reformation, it no longer suited the English Church. After Henry VIII's acrimonious break with Rome, the newly established Protestant Church of England looked back into history to provide it with the historical legitimacy to set itself apart from Roman Catholicism. To do this, scholars seized on a remark made in the sixth century by Gildas in *The Ruin of Britain* that, in what became England, the original Britons had been completely wiped out by the Saxons. The natural conclusion was that the English were the linear descendants of the Saxons, not the Britons at all. This was an undiluted and direct genealogical connection, not with the defeated Britons and the mythical Arthur, but with the victorious Saxons. In this version of events the Saxons were not the malicious and unprincipled opportunists whose foothold in Britain came about only through Vortigern's treachery. Far from it: the Saxons were strong, self-confident and adventurous pioneers who had triumphed against the weak-willed Britons through the intrinsic superiority of their moral character and their love of freedom. The English Church no longer looked west and north to the mountains of Wales and Scotland for its natural affiliations, but across the North Sea to the Teutonic Germans whose stout spirit of Protestant independence had triumphed against the corruption of the Roman Church.

To recreate the myth of an Anglo-Saxon golden age before the Norman Conquest, Protestant historians needed a hero to replace Arthur. They found one in King Alfred, and the PR campaign began: 'the great and singular qualities in this king, worthy of high renown and commendation – godly and excellent virtues, joined with a public and tender care, and a zealous study for the common peace and tranquillity . . . his heroical properties jointed together in one piece', wrote John Foxe in 1563. It clearly worked: even today, Alfred is the one Saxon king that most children have heard of – even if all they remember is that he burnt the cakes. Unlike Arthur, there is no doubt that Alfred existed, but how close the glowing tributes to both his military genius and his humble and scholarly character are to reality is still an open question. He reigned from 871 to 899 and was, as we shall see later, instrumental in preventing the Danish Vikings from over-running the whole country.

At the same time that Alfred was being resurrected in

England, Protestant scholars, including Martin Luther in Germany, were creating their own origin myths for the same reason. To reinforce their independence from the Catholic Church, they drew heavily on classical writers for their justification. One of these was the Roman historian Tacitus, who wrote in AD 98, 'For myself I accept the view that the people of Germany have never been tainted by intermarriage with other peoples and stand out as a nation peculiar, pure and unique of its kind.' Luther himself even managed to concoct a genealogy for the Germans right back to Adam, who for Christians like Luther was the father of the human race.

What began as a declaration of religious independence from Rome transformed over the years into a virulent doctrine of Saxon/Teutonic racial superiority over the other inhabitants of the Isles that has had immense and far-reaching political and social consequences. The reinvention of a glorious English past gathered pace. The Magna Carta, in essence an unimportant concordat between King John and his Norman barons, was reborn as a declaration of Saxon independence every bit as important to the English as the US Bill of Rights is to Americans. The Puritans appealed to the myth in their bitter struggle with the Crown during the English Civil War when John Hare, one of the leaders of the Parliamentarians, wrote about his side in 1640, the first year of the war:

> our progenitors that transplanted themselves from Germany hither did not commixe themselves with the ancient inhabitants of the country of the Britain's, but totally expelling them, they took the sole possession of the land to themselves, thereby preserving their blood, laws and language uncorrupted ...

Gradually the monarchy changed allegiance to suit the new origin myth. James VI/I even switched sides during the course of his own reign. Having at first asserted his entitlement to rule over both Scotland and England, based on his claim to be Merlin's Arthur reborn, he very soon afterwards basked in the appellation of the 'chiefest Blood-Royal of our ancient English-Saxon kings', according to a dedication in the influential book *Restitution of Decayed Intelligence*, written in 1605 by Richard Verstegen.

In the context of the genetics we will come to later, Verstegen was the first author to point out the potential embarrassment that the purity of the Saxon line must surely have been 'diluted' or 'contaminated' by the later arrival of large numbers of Danes and Normans. He countered this by claiming, first, that their numerical contribution was slight and, second, that both the Danes and the Normans were themselves of Germanic origin anyway, so they could have no effect on the essential racial purity of the Teutonic English.

As the myth gained momentum, the voices raised against it became fewer and further between. The writer Daniel Defoe was one exception, parodying the whole idea of English racial purity and superiority in his poem 'The True-Born Englishman', written in 1701:

> The Romans first with Julius Caesar cáme
> Including all the Nations of that Name
> Gauls, Greeks and Lombards; and by Computation
> Auxiliaries or slaves of ev'ry Nation
> With Hengist, Saxons; Danes with Sueno came
> In search of Plunder, not in search of Fame
> Scots, Picts and Irish from the Hibernian shore:
> And Conquering William brought the Normans O're.

> All these their Barb'rous offspring left behind
> the dregs of Armies, they of all Mankind;
> Blended with Britons, who before were here,
> of whom the Welch ha'blest the Character.
> From the amphibious Ill-born Mob began
> That vain ill-natured thing, an Englishman.

But dissenting voices were definitely in the minority, and the myth grew and grew, finding a new outlet in the development of the African slave trade. Though European attitudes to black people and a readiness to exploit them for personal gain were nothing new, the resurgent racial pride which accompanied the growth of the Teutonic myth encouraged further victimization. While there may have been some uncertainty about the purity of the Saxon pedigree within England, there could be no cause for doubt that black Africans had no claims whatsoever to the Teutonic blood-line with its attendant virtues of enterprise, independence and high moral character.

During the eighteenth century, the myth had grown to such prominence that it was scarcely, if ever, questioned. It attained an invincibility equal to that of the Arthurian legends of Geoffrey's *History* 500 years earlier. But now its effects were felt not just in Britain, but throughout the world. The influential French political philosopher Baron de Montesquieu wrote in 1734 that the English political system came straight from the forests of Germany, imported and elaborated by their Saxon descendants. Even the great Scottish philosopher David Hume, who constantly required evidence as the foundation for any belief, accepted without question the purity of the German race first expressed so long ago by Tacitus. Thomas Jefferson, one of the draftsmen of the Declaration of Independence, who became the third President of the United States, wrote in 1774 that it was the Saxon ancestry of the American colonists that gave them a natural right to build for themselves a free and independent state, liberated from British colonial rule.

The triumph of the Teutonic myth was almost complete as its popularity reached its peak during the nineteenth century. Indeed the superiority and self-belief with which its adherents cloaked themselves was central to the construction and administration of the British Empire. The myth gave to the Englishmen abroad the absolute conviction that their ancient Saxon pedigree imbued them with inherited qualities of honour and leadership, and the political institutions to go with it, that were far superior to any in the world. Bolstered with that ingrained sense of destiny the English did, indeed, rule the world – for a while.

But the triumph of the myth came at a price. The growing sense of racial superiority among the English set them increasingly at odds with the other inhabitants of the Isles, the Welsh and the Scots on the British mainland, and with the Irish. The simple racism of the myth collapsed all three into the same denomination, 'the Celts', and poured scorn on them.

By now, science had been harnessed to the myth in an enthusiastic attempt to build a solid frame to underpin its more extravagant assumptions. And when science and racism are mixed, the cocktail becomes increasingly volatile. At the end of his rambling book *The Races of Men*, published in 1850, Robert Knox MD, surgeon and enthusiast for the new science of comparative anatomy, concludes after 350 pages of Saxon worship and Celtic insult that, 'The Celtic Race must be forced from this soil. England's safety requires it'. This outrageous suggestion, as it appears to us now, was completely in tune with the prevailing view, if not of an actual genocide, then certainly of cultural and spiritual suppression. In a superbly argued defence of the value of Celtic literature, published in 1867, the literary critic Matthew Arnold quotes a leader from *The Times* on the subject of the Welsh language:

> The Welsh language is the curse of Wales. Its prevalence, and the ignorance of English have excluded, and even now exclude, the Welsh people from the civilisation of their English neighbours. An Eisteddfod [the annual Welsh literary and musical festival] is one of the most mischievous and selfish pieces of sentimentalism which could possibly be perpetrated. It is simply a foolish interference with the natural progress of civilisation and prosperity. If it is desirable that the Welsh should talk English, it is monstrous folly to encourage them in a loving fondness for their old language. Not only the energy and power, but the intelligence and music of Europe have come mainly from Teutonic sources, and this glorification of everything Celtic, if it were not pedantry would be sheer ignorance. The sooner all Welsh specialists disappear from the face of the earth, the better.

Even taking into account the often strident and provocative language of a *Times* leader, it is a chilling piece.

The decline of the myth's supremacy came as the nineteenth century drew to a close. In a parallel with the undermining of the factual basis of Arthurian legend and the ancient succession of British kings recounted in Geoffrey's *History*, the absolute belief in the Teutonic myth suffered a similar fate. There was no single scholar assassin like Polydore Vergil, but rather a series of snipers. One of these, the literary critic J. M. Robertson, concluded that, far from being the heralds of a superior race honed to perfection in the forests of Germany, the first Saxons were 'pagan,

non-literate and barbaric, heroes of a northern society so dis-
organized that they had little concept of national, racial or
political loyalties'.

But it was political developments in Germany, rather than
Britain, that finally sealed the fate of the Teutonic myth. Not
surprisingly, Germans also favoured the good light the myth cast
on their own racial purity and superiority with the almost
genetically linked qualities of freedom and independence. Also
keen to distance themselves from Rome, German scholars had
worked in a parallel effort to reinforce their independence with a
racially based justification. It was in the late nineteenth century
that the German origin myth first became firmly attached to the
concept of the Aryan.

The creator of the Aryan myth was a German linguist,
Max Müller, working in Oxford as a Professor of Modern
Languages. Müller did more than anyone to create this myth by
falling into the trap of unquestioningly conflating language with
race – a temptation which even contemporary scholars often seem
quite unable to resist. As a linguist, Müller was very well aware of
the similarities of the major European languages to Sanskrit and
Persian. This similarity hinted at a common origin and led
directly to the concept, widely accepted today, of an
'Indo-European' language family. The language family was
originally known as 'Aryan', from the Sanskrit word meaning
'noble'. Müller took the natural but unjustified step of mutating
this theory of language to a theory of race. He concluded that
there must have been, as well as an original Aryan language, an
original Aryan people. That most promiscuous and malignant of
racial myths was born. As his career progressed, Müller came to
doubt his invention and by 1888, to his credit, he positively
rejected it. But it was too late. The genie was out of the bottle.

As we have seen throughout this chapter, the career of a myth depends far less on its factual accuracy than on its congruence with contemporary political ambition, and the fervour with which people believe it. Towards the end of the nineteenth century, Germany, through its Chancellor Bismarck and later under Kaiser Wilhelm, recruited the Aryan myth, with its closely linked connotations of German racial superiority, to justify its own campaign of imperial expansion. These ambitions soon became a direct threat to the British Empire, also fuelled by the same myth. Enthusiasm among the British for the close affiliation with Germany that was so much part of the Teutonic myth rapidly dwindled as the two countries became enemies. Not surprisingly, after the First World War it vanished completely.

The growing distaste of the British did nothing to halt the rise and rise of the Aryan myth in Germany itself. The Nazis, now seeing themselves the sole inheritors of Aryan racial superiority, exploited it ruthlessly against their 'enemies' both within and without. Nothing underlines the dreadful power and the dreadful danger of racial myth more than the smoke rising from the chimneys of Belsen and Dachau, Treblinka and Auschwitz.

3

THE RESURGENT CELTS

As I write, the 2006 World Cup Finals draw closer and one of the few remaining expressions of English patriotic nationalism is beginning to show itself. The red cross of St George on the white background of the English flag is seen hanging from first-floor windows and fluttering from the rear windows of speeding cars. Supermarkets have 'Come on England' signs hanging above the aisles. If the 2004 European Cup Finals are anything to go by, the flags will be hastily taken down as soon as England are knocked out of the competition. Even the appointment of a foreign manager to the English football team is received with only mild remonstrations. That, and the enthusiasm surrounding the victory over Australia in the Ashes test series in the 2005 cricket season, is about all you are likely to see these days. The English national day, St George's Day – 23 April – is barely celebrated. The new ethnic myth is not to be found on the Saxon streets of London, but in the Celtic west. On St Patrick's Day – 17 March –

the streets of Dublin and of New York are packed with parades and partygoers. While the Teutonic myth has submerged beneath the surface, if only for the time being, the Celtic myth grows stronger as each year passes.

Visit any of the multitude of tourist gift shops in Ireland or the west of Scotland and you are immediately confronted by what is best described as the Celtic brand. Silver brooches with naturalistic intertwining tendrils; amethysts set in the centre of 'Celtic' crosses, the arms embossed with intricate 'Celtic' knot-work; reproductions of fabulously illustrated early Christian illuminated manuscripts. Though they are often imported from China, these are a tangible part of the material expression of the Celt, one that is recruited to market this part of the Isles through-out the world. It is also a brand that is understood by local people and expressed particularly strongly in music and, in a different way, in sport. One of the largest music festivals in Scotland is called 'Celtic Connections'. Home-grown bands subscribe in many ways, even in their names. Runrig, a very successful band from Skye off the north-west coast of Scotland, is named after an old land-usage system and recalls the frugal life of peasant farmers. In sport, football teams declare their links to the myth and none more so than the world-famous Glasgow side Celtic, locked in perpetual and often bitter rivalry with Glasgow Rangers. In rugby there is a vigorous Celtic League comprising teams from Scotland, Wales and Ireland.

The brand is supported by the cultural glue of the Gaelic language, which binds the west of Scotland with Ireland and, in a slightly different form, with Wales, Cornwall and Brittany. But perhaps the dominant feature of the Celtic brand is that it joins the west together and deliberately separates it from the rest of the Isles and the perceived domination of the

English. It is important for the modern Celt to be *different*.

Even though Celtishness is today mainly expressed in language, music, sport and other cultural pursuits, there lurks beneath it an unspoken belief in some form of ancient Celtic race whose descendants live on today. Could genetics test this assumption? Is there a genetic basis for this underlying belief in a race, or races, of ancient Celts and can we show it by sifting through the genes of today's 'Celts'? Or is Celtishness a purely cultural phenomenon, at once sincerely felt and eagerly exploited but with no underlying biological framework?

If behind the paraphernalia of the Celtic brand there really does lie some grain of substance in the notion of a Celtic people, this immediately begs the question of when they arrived in the Isles and where they came from. Indeed, where does the notion that the Celts ever existed as a separate people, capable of acting together, moving together and arriving somewhere, actually stem from? The notion, oddly enough, is a surprisingly recent one. It began to take shape in the years around 1700 when Edward Lhuyd, from Oswestry on the Welsh border, became the director of the Ashmolean Museum in Oxford. Lhuyd travelled widely in Ireland, Wales and the Scottish Highlands, collecting antiquities and manuscripts for the museum and recording the folklore of the lands he visited. On his travels he noticed the similarities between Welsh, Cornish, Breton, Irish and Scots Gaelic and the ancient languages of Gaul. In his book *Archaeologia Britannica*, published in 1707, he was the first to group these languages together and embrace them under the generic term of Celtic. He was also the first to point out that the languages belonged to two distinct sets, distinguished from each other by their pronunciation. The harsher consonants of Breton, Cornish and Welsh (as in *ap*, meaning 'son of') led Lhuyd to call these the P-Celtic languages, while the softer sounds of Irish and Scots Gaelic (as in *mac* with the same meaning) were referred to by Lhuyd as Q-Celtic. Having found a language family, it was all too easy to invent a people and Lhuyd very soon constructed a historical explanation of how this linguistic continuity may have come about. He suggested that, first of all, Irish Britons moved to the Isles, but were pushed into Scotland and northern Britain by a second wave of Gauls from France, who then occupied Wales and the south and west of England.

Implicit in all of this is the concept that there existed one or more groups of Celts who moved around from one place to another, taking their language with them as they went. This is an idea in the grand tradition of migration as the sole explanation for cultural change – a tradition which until recently dominated not only linguistics but archaeology as well. A type of pottery or a particular burial ritual found in two different places was taken as proof that people from one moved to occupy the second. This type of reasoning drove archaeology for most of the twentieth century and became the standard dogma for the spread of any cultural change, be it language, weapon design, stone tools or even agriculture. In the last twenty years or so the pendulum of academic fashion has begun to swing to the other extreme, where nobody actually moves anywhere except to pass on their ideas and scurry back home.

But back to the Celts. Edward Lhuyd, though he helped create the concept of the Celtic people, did not invent the word. It makes its first appearance as *Keltoi* in ancient Greek, where it is used as a derogatory catch-all name for strangers and foreigners, people from another place. Uncivilized, rough, uncouth, not 'one of us'. By the time Julius Caesar wrote his *Gallic Wars*, around 60 BC, the people of Gaul, according to Caesar, called themselves Celts. So while the Greeks used *Keltoi* to refer to outsiders, coming from beyond the limits of the civilized Mediterranean world, the name itself might originally have come from one or more of the tribes themselves. For the Romans, the terms Celt and Gaul were pretty much interchangeable, used to describe the inhabitants of their territories in France and northern Italy and to tell them apart from the real enemy – the Germans.

However, when we come to the people of Britain and Ireland during the Roman period, nobody called them Celts. They called them a lot of things, but not Celts. Neither is there any record of anyone from the Isles using the word Celt to describe themselves until the eighteenth century, after Edward Lhuyd had reinvented the term for his language family and then for the people who spoke it. If a Celt is someone who speaks one of the Celtic languages as defined by Lhuyd, then everyone in Britain and Ireland would have been a Celt when the Romans invaded. If a Celt is someone whose ancestors lived in the parts of the Isles where these languages are still spoken today, then the definition becomes much narrower and more akin to what the Celtic brand now represents. So, just as to the Greeks, there is no precise definition of Celt. It is amorphous, fluid, capable of many simultaneous meanings. To some, like the archaeologist Simon James, it is a shameless invention. In his angry polemic *The Atlantic Celts: Ancient People or Modern Invention?*, James concludes that 'The ancient Celts are an essentially bogus and recent invention', an invention used most recently for political purposes in the lead-up to Scottish and Welsh devolution. C. S. Lewis expressed the ambiguity and uncertainty in softer tones when he wrote, 'anything is possible in the fabulous Celtic twilight, which is not so much a twilight of the Gods but of reason'.

When it comes to getting hold of a definition of the Celt, or

Celtic, a definition to be tested by genetics, I found myself struggling, enveloped in a mist of uncertainty and enigma. For sure there was the marketable expression of Celticity, the silver brooches, the tartan ties, the kilts. But these are caricatures of something much deeper. What it means to be Celtic, to feel Celtic, is very different. As is to be expected of him, Sir Walter Scott's description of the Celtic Muse is highly sentimental. He writes in his novel *Waverley*:

> To speak in the poetical language of my country, the seat of the Celtic muse is in the mist of the secret and solitary hill, and her voice is the murmur of the mountain stream. He who wooes her must love the barren rock more than the fertile valley and the solitude of the desert better than the festivity of the hall.

And yet there is something in what Scott says. The emotional, almost the physical, attachment to the land is central to the poetry of the Celt. Out of term time, when I am not required to be in Oxford, I live on the Isle of Skye. My house once belonged to Sorley Maclean, widely acclaimed as the greatest Gaelic poet of the twentieth century. In fact, that is where I am writing this chapter and it is in his old filing cabinet that the manuscript will remain until I send it off to be typed. Sorley's poetry is rich in reference to the woods, the sea and the hills. In 'The Cuillin' he writes:

> Loch of loches in Coire Lagain
> Were it not for the springs of Coire Mhadaidh
> The spring above all other springs
> In the green and white Fair Corrie.

Coire Lagain is a high place in the Cuillin Hills of Skye, hemmed in by hundreds of feet of the steep rock ramparts that protect the high ridge. But this is not at all a romanticized description of the land, for the poem goes on to another familiar theme which permeates the culture of the Highlands – that of loss and unquestionable sadness.

> Multitude of springs and fewness of young men
> today, yesterday and last night keeping me awake:
> the miserable loss of our country's people
> clearing of tenants, exile, exploitation
> and the great island is seen with its winding shores
> a hoodie-crow squatting on each dun
> black soft squinting hoodie-crows
> who think themselves all eagles.

The loss which Sorley mourns in this and other poems is at once the people forced to leave their homes during the notorious Highland Clearances in the late eighteenth and early nineteenth centuries and also the language, Gaelic, which was the language of his poems. At first he wrote in Gaelic and English but in 1933, when he was twenty-two, he decided to write only in Gaelic and he destroyed those of his English works that he could lay his hands on.

Gaelic and her cousin tongues are a strong unifying force of the Celtic lands. Their fortunes, in Scotland, in Ireland, in Wales and also in Brittany, are a barometer of the self-confidence of the people who call themselves Celts. Since Celtic was a linguistic definition in the first place, this seems only appropriate.

In Skye, as in many parts of the Highlands, there is a palpable sense of a Gaelic revival, a renaissance in poetry and music and above all in the language. The steady decline in Gaelic speakers – it is spoken as a first tongue by only a few thousand people in the Hebrides, most of them in middle age or beyond – has been halted by the welcome introduction in 1986, after decades of lobbying, of Gaelic-medium education in primary schools, where all lessons are given in that language. Most children whose parents have the choice opt for lessons in the Gaelic stream rather than the English alternative. Now Skye children can go right the way through school being taught in Gaelic and, in recent years, go on to tertiary education in Gaelic at the world's first Gaelic College at Sabhal Mor Ostaig in Sleat on the southernmost of Skye's many peninsulas. Whether this very hard-fought initiative will reverse the decline in the language in the long term remains to be seen, but I have never visited a higher education institute anywhere in the world that is so brimming with confidence and enthusiasm for its mission in life.

Sabhal Mor kindly allowed me to use their library for my research – a library with what must be the best view in the world. Sabhal Mor (pronounced Sall More and meaning simply 'Big Barn' in Gaelic) is perched on a promontory overlooking the Sound of Sleat; the view takes in the distant outline of Ardnamurchan and the sands of Morar to the south and up to the hills above Glenelg and Kyle of Lochalsh to the north. But straight ahead, across 3 miles of blue and wind-blown sea, are the mountains of Knoydart, yellow and brown in the autumn setting sun. Knoydart, between the secret lochs of Nevis and Hiourn, was once a prosperous community of twenty-seven crofting townships and 3,500 people. Now it is empty, save for a cluster of white houses I can see on the shore at Airor. The Knoydart estate was cleared of people in the 1840s by the landlord, Sir Ranald McDonnell of Glengarry, to make way for the more profitable sheep. This is an all too familiar story in the Highlands, though nowhere was as thoroughly cleansed as Knoydart, and it has been vividly recounted many times. Here is one from Neil Gunn, a Scottish novelist of the early twentieth century:

> As always the recollection is dominated by dramatic images – the ragged remnants of a once proud peasantry hounded from the hills by the factors and police were driven aboard disease-ridden ships bound for outlandish colonies, their families broken, their ministers compliant and the collective agony sounded by the pibroch and the wailing of pathetic humanity.

By and large, the English were blamed for this human translocation and spiritual genocide, not that the landlords were themselves English but came from a heavily anglicized Scottish aristocracy who spent most of their time in London. Still the Celtic identity, in Ireland, Wales and Scotland, and the language, defines itself in part at least as being 'not English'. That is not to say it is an aggressive demarcation, and as an Englishman with very little Gaelic living on Skye I have never been made to feel less than welcome.

Of course, the main emigration of the late eighteenth and early nineteenth centuries, whether by the forced hand of the landlord or for the opportunities for a better life on offer in Glasgow and the other industrial towns of the Central Belt or the colonies, meant that as people left Scotland, and Ireland too, they arrived somewhere else. Estimates vary, but one set of figures has it that there are 28 million people of Scottish and 16 million of Irish descent spread throughout the world. Even if these figures are way off the mark, and they are conservative estimates, there are now far more Celts living overseas than in the Isles. Most made their homes in the New World, mainly the USA and Canada, but emigration to Australia, New Zealand and to a lesser extent South Africa adds millions to this list. In some places, like the southern part of South Island, New Zealand, the Scots practically took over the whole country and, tellingly, the principal town Dunedin has the Gaelic name for Edinburgh.

Many emigrating Celts and their descendants did extremely well, of course, but ironically, given the circumstances of their leaving, they were also sometimes guilty of dispossessing the indigenous people of their tribal lands. As Paul Besu, a social anthropologist from University College London, writes:

Scots pioneers in Victoria (Australia) were often land-grabbers and squatters who were notorious for their ruthlessness and the Scots, like the English, Welsh and Irish, played a full part in the harsh treatment of Aboriginal peoples. It was ironic that some of the most notoriously involved were Highlanders who themselves had suffered clearance and privation in the old country.

Paul Besu was researching what it was that drew the descendants of these emigrants to search out their roots in Scotland and he interviewed people about their reasons for making these journeys from the other side of the world. Tens of thousands of Americans, Canadians, Australians and New Zealanders come to Scotland and Ireland every year to seek out and placate their innermost desires to see and feel the homelands of their ancestors. Of course there are comparable numbers of visitors on similar missions to England too, to reconnect, but theirs is a slightly different mission, perhaps less romantic and more matter of fact. The general causes of their emigration do not usually include being driven from the land.

I have experienced the thirst for roots first hand through the company I set up to help people trace their origins using DNA. We have thousands of customers, many from precisely those locations that once received Scots and Irish emigrants. Often a DNA test will accompany a journey to the homeland and even when it does not, a DNA test which roots a person to Scotland or Ireland makes a living link between descendant and ancestor. It is all the more powerful as this talisman is carried across the generations in every cell of the body, as it was in the bodies of ancestors, including the ones who made the journeys 'aboard disease-ridden ships bound for outlandish colonies'. It was there.

There have been, as you may have suspected, plenty of theories about what draws people to search out their roots. Behind the sociology-speak, such as

> the contemporary quest for roots is a response to the trauma of displacement associated with migration which has become a global commonplace and individuals are able to conduct meaningful, morally defensible and authentic self-narratives from the ambiguities and discontinuities of their migrant histories, thus recovering a sense of being 'at home' in the 'maelstrom' of modernity . . .

are much pithier and more articulate reasons, as revealed in Paul Besu's survey. An anonymous Australian from New South Wales simply said, 'I want to be able to tell my children where their ancestors came from. It gives them a sense of belonging in a world that sometimes moves too fast.'

Another Australian, improbably called Anne Roots, told Besu:

> I am a fourth-generation Australian but I know that the thread reaching back to the obscure past has never been broken. The

process of evolution has failed to break the translucent thread that is mysteriously joined to the Isle of Skye. I cannot explain some of my experiences, or why I wanted to go to the Hebrides before I knew some of my forebears came from there. My only explanation is that the spirit of my ancestors kept calling me back.

Janet, from Geelong, south of Melbourne, made the fascinating comparison when she visited her ancestral home of Paabay, an island in the Sound of Harris, that 'to my mind the Celt, in a British context, is to the Anglo-Saxon what the Aboriginal, in an Australian context, is to the settler.'

The intense spirituality of the Australian Aborigine, the connection to ancestors and the homeland, is in a muted form reflected in the search for Celtic roots. Displaced by the invader and forced to the margins before being forced into exile overseas, the Celt is perceived to be the British – or even the European – aboriginal. She continues, 'to have Celtic roots is to demonstrate that one also has a rich, tribal heritage rooted deeply within a landscape that is both mystical and mythical'.

And it is the case that in 'New Age' bookshops around the world, titles on Celtic spirituality are found on the same shelves as Aboriginal and Native American material in the same genre. However, be warned that I heard the distinguished American sociologist Michael Waltzer in a recent lecture dismiss excess spirituality as 'the solace of a conquered people'.

Before we move on to more solid ground, let me just mention Frank from Boulder, Colorado. After spending twelve years with Native American teachers, Frank took part in the Sun Dance ceremony of the Lakota people, an experience which set him on the path to discovering his Celtic heritage. He now describes himself as 'a poet, ecopsychologist and visionary teacher in the Celtic spiritual tradition'. Frank leads pilgrimages to the Scottish Highlands to promote what he calls 'Highland cultural soul retrieval'.

The range of emotion covered by the Celtic umbrella is vast, from a feeling of displacement and affinity with aboriginal groups, to a successful marketing tool, to a political rallying call, to the focus for sporting identity, even fanaticism. Can genetics lift the veil and see what lies beneath? Faced with this multiplicity of meaning for Celt and Celtic, what range of possibilities should we expect genetics to reveal? Might we be able to detect the waves of a large-scale migration envisaged by Edward Lhuyd? Or might we find evidence that what we now call Celts have been here all along? Will we find any genetic similarity between the present-day Celts and the people of the rest of Britain, or will there be a sharp divide? And where should we look for origins? Though not absolutely essential for success in historical genetics, it is always best to formulate some scenarios that can be tested.

One of the most striking emblems of the Celtic brand, the intricate naturalistic knotwork that inspires the modern Celtic jeweller, had its origin not in the Atlantic communities linked by a common language, but in central Europe. The evolution of this highly distinctive art form coincided with the rise of rich settlements north of the Alps, centres which controlled the trade of goods like amber and tin, flowing south to the Mediterranean world and their exchange for luxuries, such as wine and jewellery. In all likelihood, these luxury imports were used by local chieftains as a badge of status and also distributed among their subordinates in exchange for favours and services.

The trading settlements spanned the heartland of Europe where its great navigable rivers converge in a relatively small area in eastern France and Switzerland. The Loire going westward to

the Atlantic, the Rhône south to the Mediterranean, the Rhine
north to the North Sea and the Danube east to the Black Sea.
These were the arteries of prehistoric Europe along which flowed
the life-blood of trade. Whoever controlled the heads of the rivers
and the land between them controlled the trade – and grew very
rich on it. At the peak, around 600 BC, there was enough wealth
to stimulate and support the production of a local style of craft-
work, and this is where we see the first appearance, principally in
the delicate metalwork, of what we now call Celtic Art. The La
Tène style, which we now most strongly associate with the Celtic
brand, began not on the ocean coasts of the Atlantic, but within
sight of the Alps.

But was it just the goods and the ideas that moved, or was it the
people migrating *en masse* from central Europe to the far west?
Although there is very good archaeological and historical evidence
that people from this region did indeed move in numbers east and
south to Greece, where they attacked the temple at Delphi in 273 BC,
before finally settling in central Turkey, there is no evidence at all
that the ancestors of today's Celts of the Isles took the opposite track
and ended up in Britain. Yet, although support for the popular
notion that the Celtic people of the Isles travelled across land from
central Europe may be entirely lacking, we may still find the
evidence for it in the genetics.

However, the most obvious of routes linking today's Celts of
the Isles is not the land at all but the sea. Motorways and fast roads
have inverted in our minds the comparative difficulty of moving
across land and water. In ancient times, and indeed until the last
two centuries, getting around by boat was a lot easier than
travelling over the land. Until the rise of, first, the railway and
then the car and the lorry, water was the way to travel. Was a sea
route to the Isles the more likely?

At school we are taught that 'civilization' arose around the Mediterranean, in the ancient cities of Egypt, and that we trace the origins of our culture and our political processes to the countries bordering that almost landlocked sea. Our taught impression of life beyond the Strait of Gibraltar is one of barbarism and savagery, rather like the Greeks' view of the *Keltoi*. We are taught nothing of the vigorous culture and the technological achievements of the Atlantic seaboard, the coastline stretching from North Africa in the south 2,000 miles to Shetland off the north coast of Scotland and beyond to Scandinavia. But this Atlantic zone has a prehistory as ancient and as colourful as any in the Mediterranean. There were people living along this coastline 8,000 years ago and they were using boats not just for cruising close to the shore but for venturing out into deep water, judging by the types of fish whose remains litter their encampments. None of these sea-going vessels survives, which is no surprise since they would have been made of perishable wood and animal skins. By 6,000 years ago, agriculture had seeped into the region via the Mediterranean coastline, evidence once again of the maritime traffic. The first, literally, hard evidence of widespread exchanges along the coast came in the form of distinctive polished stone axes, manufactured in Brittany, which found their way all along the coast of France and Spain to the south, and north across the sea to Cornwall. But the most dramatic examples of continuity along the Atlantic zone are the great stone monuments, the megaliths, which rise from the ground from Orkney and Lewis in the north to Spain and Portugal in the south. These are a purely Atlantic phenomenon, owing nothing at all to the Mediterranean world. Could it be that it was by this route that the Celts of the Isles first arrived?

4

THE SKULL SNATCHERS

The first forays of science into the highly charged arguments about British origins came at the height of the Victorian enthusiasm for Saxon superiority. It is hard to imagine how ingrained was the sense that the people of Britain were split into two entirely different 'races' and how superior the Saxons felt about themselves. Just to remind us, I quote again from the extremely popular if eccentric author, the surgeon Robert Knox. He wrote that 'Race is everything, literature, science, art – in a word, civilisation depends on it'. And Knox left his readers in no doubt where his sympathies lay in the debate on the racial character of Celt and Saxon. The Saxon, he claims, 'cannot sit still an instant, so powerful is the desire for work, labour, excitement, muscular exertion'. The Celts, on the other hand – judging by such woodcut illustrations as 'A Celtic groupe, such as may be seen at any time in Marylebone, London', in which a group of deformed and decidedly dodgy characters glower from the page – are the complete opposite: irredeemable malingerers.

The text is no more flattering. On the notorious Highland Clearances, he writes: 'the dreamy Celt exclaims at the parting moment from the horrid land of his birth "we'll maybe return to Lochaber no more." And why should you return, miserable and wretched man, to the dark and filthy hovel you never sought to purify?'

Knox pulls out all the stops when it comes to the Celts of Ireland: 'the source of all evil lies in the race, the Celtic race of Ireland. The race must be forced from the soil, by fair means if possible, still they must leave'. A few sentences later is an entreaty to genocide no less chilling in intent than in Bosnia or Rwanda:

> The Orange Club of Ireland [an extreme protestant group] is a Saxon confederation for the clearing of the land of all papists and jacobites; this means Celts. If left to themselves they would clear them out, as Cromwell proposed, by the sword; it would not require six weeks to accomplish this work.

By the time Knox was penning his poisonous invective, in the mid-nineteenth century, science was making itself felt in all walks of life. The appeal to rational arbitration of such issues as the racial purity, or otherwise, of Celt and Saxon had obvious attractions to those with a more liberal outlook than the likes of Robert Knox. The most articulate of these, Matthew Arnold, literary critic and a prominent champion of Celtic literature, despaired of the wedge being driven between Celt and Saxon, not just by fanatics like Knox, but by powerful and influential members of the British Establishment. Men like Lord Lyndhurst, whose description of the Irish as 'Aliens in speech, in religion, in blood, makes the estrangement [of Celtic and Saxon] immense, incurable, fatal'. Feeling forced to react, Matthew Arnold makes an optimistic appeal: 'Fanciful as this notion may seem, I am inclined to think that the march of science will insist that there is no such original chasm between the Celt and the Saxon as we once popularly imagined'.

But what was the basis for Matthew Arnold's optimism? It was this. That even if many thousands or hundreds of thousands of Saxons arrived in the centuries following Hengist, they would, within a few generations, intermarry and blend with the Britons already here. This was anathema to racial purists – it just could not happen. Races were pure and indivisible. But how could this theory of racial purity overcome the all too apparent empirical fact – especially obvious as white imperial boundaries expanded into Africa, India and America – that there were no barriers to mating between 'races'? The answer came that the offspring of such matings were weakened hybrids, incapable of sustaining themselves over more than a few generations. How this worked in practice was explained using the Spanish 'conquest' of South America and the interbreeding which followed. According to Knox:

When the best blood in Spain migrated to America, they killed as many of the natives as they could. But this could not go on, labourers to till the soil being required. Then came the admixture with the Indian blood and the Iberian blood, the produce being the mulatto.

Even the name, Spanish for 'little mule', recalls the sterile hybrid of horse and donkey. Knox continues:

as a hybrid he [the mulatto] becomes non-productive after a time, if he intermarries only with the mulatto. Thus, year by year, the Spanish blood disappears, and with it the mulatto, and the population, retrograding towards the indigenous inhabitants, returns to that Indian population, the hereditary descendants of those whom Cortes found there.

Races, in this exposition, do not hybridize and any unnatural mixing produces only enfeebled offspring whose progeny are doomed to extinction. Though the nineteenth century was dominated by the extreme views of people like Robert Knox, who believed in the sanctity and purity of racial groups – with Saxons at the top of the rankings, of course – there were a few lone voices raised against the predominant dogma. One of these was Luke Owen Pike, a Lincoln's Inn barrister. His well-argued, and witty, riposte to the Teutomaniacs like Knox was to point out that it was extremely unlikely, even if the entire population of Jutes, Angles and Saxons arrived in Britain, that they could have exterminated *all* the Britons, with their centuries of experience of Roman military tactics. Even if they had managed to kill all the men, they would not have killed all the women.

The women and the children, at least, are doomed to a different, if not a happier fate. And for this reason it must almost always happen that, after the conquest of any country, the blood of the original inhabitants will still preponderate. There is no reason to suppose that the result was different in the case of the Saxon conquest.

Pike not only rejected the concept of the immiscibility of races, he argued for the creation of a hybrid racial mixture in which the indigenous component would usually predominate.

Although the ranting racist diatribes of Robert Knox, the moderation of Luke Owen Pike, even the commentaries of Matthew Arnold, were the expression of strongly held opinions, none of them had a solid basis of factual evidence. While the fierce argument was raging about whether races were fixed and immiscible or could happily and successfully interbreed and blend, a few people did begin to gather systematic scientific observations to inform the debate.

The first to do so on a significant scale was John Beddoe, a doctor who spent the best part of his life travelling to every part of Britain recording the physical appearance of the natives, both alive and dead. He was a classic case of the Victorian amateur scholar, amassing a huge amount of data which, in sheer bulk alone, has never been surpassed. John Beddoe was born in 1826 in rural Worcestershire, the second of eight children. Though his family was comfortably well off, John was a sickly child and missed most of his formal education. Nevertheless he managed, through family connections, to get a place at University College London to study medicine. He eventually graduated, not in London but in Edinburgh, and after a spell in the Crimea set himself up in Bristol. Building up his medical practice in the fashionable quarter of Clifton was difficult, especially as he had to compete for patients with a resident pool of extremely competent and well-established doctors. With time on his hands, he began to indulge his passion for observing and recording people's appearance.

First, he had to devise a reliable classification for the features he decided to concentrate on – the colour of the hair and the colour of the eyes – exactly those features we use ourselves in the first description of a stranger. He also wanted to be quite sure that he was looking at permanent features, not something that would change from year to year. For this reason he rejected skin colour, perhaps an obvious one to include, because he was worried that it might be influenced by exposure to sunlight, which of course it is. He also decided against recording skin colour because there was a theory doing the rounds that daily exposure to smoke and grime made city-dwellers darker and darker as they got older, while their rural contemporaries remained fresh-faced and pale in comparison.

John Beddoe was determined to break free from the generalizations that were so commonplace, and still are, about regional differences in appearance. He disregarded the clichés of short, dark Welshmen or muscular, red-headed Highlanders and set out to replace these prejudiced impressions with real observations. He frequently discovered that what had been written about a place and its people was completely at odds with reality, even when the source of the misleading reports would normally have given no cause for doubt. For example, the Church of Scotland minister in Wick, a town at the north-east tip of Scotland not far from John O'Groats, was obliged to compile a statistical account of his parishioners, including their overall appearance. The minister described his flock as 'having for the most part dark brown or black hair, and dark complexions, remarkably few having red or yellow hair'. But when Beddoe arrived, he found the complete opposite. Among more than 300 individuals whose appearance he recorded, blonds and redheads were in the majority.

How did Beddoe make his observations? You can imagine how this might get very complicated – are those eyes green or hazel? Is that hair light brunette or dark blond? But Beddoe needed something much simpler, and easy to record – we will see why in a moment – and he spent several months refining his system. He decided to create just three categories of eye colour and five for hair. For eyes they were Class 1 light, Class 2 intermediate or neutral and Class 3 dark. In the light category, Class 1, were included all the blue eyes plus bluish grey, light grey and very light green. In Class 3 he put black and deep brown eyes. Class 2 included most shades of green and hazel, very light brown and very dark grey. It is not a particularly refined system, but it succeeds in its simplicity. I've tried it and I can almost always put someone's eye colour into one of the three categories at a glance without any difficulty.

When it comes to hair, though, it is harder. Most of my women friends over the age of forty probably colour their hair. Actually several have forgotten what their original hair colour was, even growing their hair out for a few weeks to be reminded, before going straight back to the hairdresser when they find out. This is nothing new and Beddoe was well aware of artificial hair colouring and its changing fashion, even in the nineteenth century. 'When I began work in England,' he wrote, 'dark hair was in fashion among the women, and light and reddish lines were dulled by greasy unguents. In later years, fair hair has been more in fashion, and golden shades, sometimes unknown to nature, are produced by art'.

In Beddoe's time, these artificial hair colours were far more confined to the wealthy than they are now. Beddoe was much more interested in the 'ordinary folk' than the 'upper classes', as he called them, as they were, in his opinion, 'more migratory and more often mixed in blood'. He eventually settled on five classes of hair colour: R for red and shades of auburn which were nearer red than brown; F for fair, including blond and very light brown hair, along with pale auburn; B included all the other shades of mid-brown; D was reserved for very dark brown; and N for the few cases of jet-black hair which he encountered.

Beddoe developed a routine. He arrived at a location and walked casually around looking at everyone who passed within 3 yards. In the palm of his left hand he held a small card divided by lines into columns and rows. In his right hand he concealed a pencil and, as people passed by, he put a tick in the appropriate square on the card. As one card filled up he replaced it, then at the end of each day worked out a simple numerical score for that locality. He called this score the 'Index of Nigrescence', which he calculated by adding the number of dark-brown (class D) scores to twice the number of jet-black (N) then subtracting the fair (F) and red (R). The ubiquitous mid-browns (B) were omitted from the calculation. He explains why he doubled the influence of the jet-blacks in the formula. It was to 'give the proper value to the greater tendency to melanosity shown thereby'. That sounds rather arbitrary to me, but the jet-blacks were so rare it didn't make a lot of difference. The simple equation for each place was:

Index of Nigrescence = D + 2N − F − R = Brown + (2 × Black) − Fair − Red

Beddoe was not under any illusion that colour of hair and eyes were excessively important features, but they did have two persuasive advantages from a practical point of view. Firstly there was no shortage of material − everyone had eyes and most had hair. The other crucial point was that there was no need to ask the subject's permission. As we shall see, Beddoe was also fascinated by the shape of people's heads, but these observations were not so straightforward. Although everyone had a head, so again there was no shortage of material, to get any sort of shape measurement he did need the subject's acquiescence, something that was un-necessary for recording hair and eye colour.

Beddoe's medical practice in Bristol slowly improved, and he did at last succeed in getting a hospital position, the equivalent of a modern-day consultant, at the Bristol Royal Infirmary. Even though this increased his workload considerably, he still found time for his travels, card and pencil in hand, to every part of the Isles. He did not restrict himself to mainland Britain, but went often to Ireland, recording four expeditions between 1860 and 1870. By now he was a well-connected physician and his excursions were rarely solitary. On his first visit to Ireland, for example, he was accompanied at various times by a Scottish archaeologist, an expert on Irish criminology, an Irish antiquarian and a Catholic priest who acted as interpreter when they needed one in the Gaelic-speaking west. The intrepid wanderers were entertained wherever they went. In Dublin they met the leading Irish physician Sir William Wilde, 'father of the unhappy Oscar', as Beddoe describes him, and an enthusiastic amateur anthropologist, who, like Beddoe, was much taken with the contemporary craze for phrenology. By this time, Beddoe was making measurements of skull shapes when he could. As we saw, this needed the acquiescence of the subject, something he obtained by the following trick.

Whenever a group of eligible peasants had collected around our party, two of us would get up a dispute as to which had the larger head, and I was called in to settle the doubt with my calipers and measuring tape. The interest of Paddy [sic] was quickly excited. Before I had finished, several of the bystanders would be wagering on the respective sizes of their own heads, and begging me to settle their differences by measurement. But such people if approached directly, always broke away at once, suspecting some concealed mischief devised by the 'Government'.

This was only a mild subterfuge compared to what happened next. Such was their enthusiasm for comparative anatomy that Beddoe and his companions turned into grave-robbers. 'The acquisition of skulls also had its difficulties,' he wrote. 'These relics lay about in old and deserted burial grounds, apparently quite uncared for, but their open abstraction would have aroused bitter feeling, and perhaps active opposition'.

To conceal what he was doing on these occasions, Beddoe wore a shooting jacket and, while his companions diverted the attention of any onlookers, he stuffed the skulls into large pockets sewn into the lining. These skulls eventually found their way to the museum of the College of Surgeons in London, where they remain to this day.

A professor from Galway developed an even more extravagant method for anatomical larceny. He always went hunting for skulls accompanied by his wife, who, in the fashion of the times, wore a wide crinoline skirt. When they spotted a skull, she stood nearby while the professor knelt down and quickly transferred the contraband to specially constructed pockets beneath the folds of his wife's voluminous skirt.

However, Beddoe's tours were not mere leisurely excursions between one literary salon and another, punctuated by a bit of grave-robbing. He wanted to get everywhere, driven by his passion not to miss a single opportunity to observe, to measure or even to steal. For example, to reach the Aran Islands in Galway Bay on the west coast of Ireland they were rowed over 10 miles of rough sea in a small skiff. Frightened and seasick, the party arrived at Inishmore, the largest of the islands, to find they had to sleep on beds of straw. They spent two days sketching the buildings and tabulating the remote population, and were even able to 'annex' a couple of skulls 'of great but unknown antiquity' from an old cemetery.

Though Beddoe's passion for collecting skulls drove him to subterfuge and theft, he did at least confine his enthusiasm to the long dead. Not so his companion, Barnard Davis, who often visited Beddoe in his Bristol practice on the look-out for interesting specimens among his friend's still-living patients. On one visit to the Infirmary, Davis was introduced to a Bosnian sailor who was desperately ill after nearly drowning when his ship sank in the Bristol Channel. He had developed gangrene in his lungs and, in the days before antibiotics, was not expected to survive. Davis, convinced the man was not much longer for this world, was unsympathetically matter of fact.

'Now,' he said to Beddoe, 'you know that man can't recover. Do take care to secure his head for me when he dies, for I have no cranium from that neighbourhood.'

Yet Davis was to be disappointed for, as Beddoe recalls, the man 'made a wonderful recovery, and carried his head back to the Adriatic on his own shoulders'.

Beddoe's reputation began to spread after he entered, and won, the Welsh National Eisteddfod competition in 1867 for the best essay on the origins of the British nation. As a young doctor struggling to get his practice established, he needed money and the annual prize of 100 guineas was a definite attraction. The prize had not been awarded for the previous four years because the entries had failed sufficiently to impress the judges. To encourage a better field, the money was raised to 150 guineas. As soon as Beddoe heard this he rapidly wrote and submitted his essay. When, to his delight, he heard that he had won, he rushed up to the Eisteddfod at Ruthin in North Wales to collect the money. He got the 100 guineas but the extra 50 never materialized. Even forty years later, when he wrote his autobiography, this episode evidently still rankled.

It took another two decades until his prize essay, plus the additional material he had acquired in the meantime, eventually appeared in book form. It was published as *The Races of Britain* in 1885. By then a Fellow of both the Royal Society and the Royal College of Physicians, the publication of *The Races of Britain* brought Beddoe fresh honours and fresh activity. In 1891 he retired from his Bristol medical practice and moved to the nearby town of Bradford-on-Avon. Even this move could not interrupt the flow of honours and invitations, and in 1905, now aged seventy-nine, he gave the annual Huxley Lecture for the Royal Anthropological Institute. He died six years later, the year after publishing his autobiography, *Memories of Eighty Years*. In the copy in front of me as I write, the sepia photograph of the frontispiece, protected by a thin sheet of tissue paper, shows an old man with a full white beard, dressed in a three-piece suit and standing confidently, legs slightly apart, on the broad front step of his stone house. One hand in his pocket, the other on a stout pole, his chained watch just visible in his waistcoat pocket, he peers into the distance with dark eyes. Eyes that had inspected and recorded the faces of thousands upon thousands of the people of the Isles. Underneath is his signature, in jet-black ink, not faded by the years. John Beddoe. The two 'd's are a little shaky, but the flourish at the end is not. It is strange to think that the hand that signed this copy, nearly 100 years ago, was the same that marked the cards he carried with him throughout his journeys.

The real meat of Beddoe's lifetime of observation lies in the tables and maps that make up about half of the 300 pages of *The Races of Britain*. He visited and recorded in 472 different locations throughout Britain and Ireland, making a total of 43,000 observations. The tables themselves are delightfully annotated with asides such as 'Cornwall, St Austell. Flower show. Country folk' or 'Bristol. Whit-Monday. Young people numerous. Dancing.' He also got hold of a further set of 13,800 observations from the unlikely source of the lists of army deserters whose pursuers published their physical description in the chillingly titled periodical *Hue and Cry*. Finally, he recorded his own patients as they came through his Bristol surgery – a total of 4,390 altogether. These were particularly precious because there was time to make accurate observations and to check on the birthplace of each patient.

Beddoe was very well aware of the dangers of unrepresentative sampling, and of more subtle influences on the accuracy of his record. For example, were his Bristol patients representative of the general healthy population? Possibly not. They must have had a reason to be in his surgery in the first place. Indeed, he notes that the incidence of disease among American army recruits was reported to be much higher among the 'dark complexioned'. And his patients did, when averaged out, have a slightly higher Index of Nigrescence than the West Country folk that he observed in streets and marketplaces. This discrepancy he puts down to differences of moral character, allying cheerfulness and an optimistic outlook to a light complexion, while 'persons of melancholic temperament (and dark complexion) I am disposed to think, resort to hospitals more frequently than the sanguine'. Even then, blondes had more fun.

The Races of Britain was, and still is, a masterpiece of observation. The samples were not statistically controlled, his coverage of Britain and Ireland was not uniform, and he came in for criticism on these grounds – rather predictably from people who never themselves got into the field. But his work is best judged as a masterly piece of natural history and not a modern work embroidered with statistical treatments.

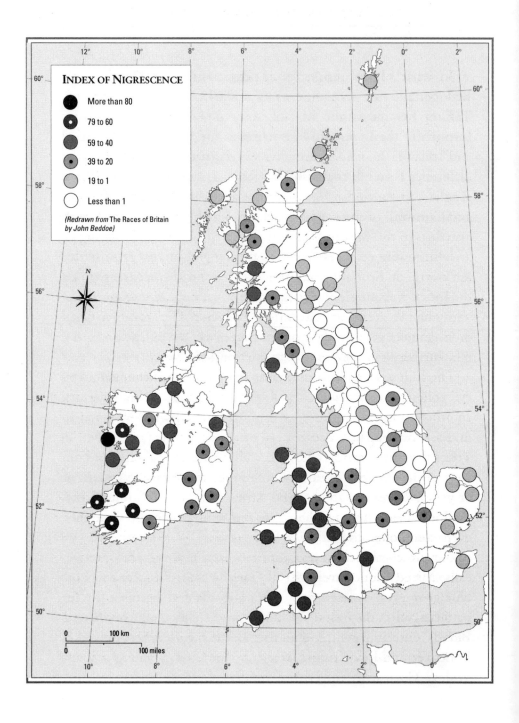

INDEX OF NIGRESCENCE

- More than 80
- 79 to 60
- 59 to 40
- 39 to 20
- 19 to 1
- Less than 1

(Redrawn from The Races of Britain by John Beddoe)

100 km

100 miles

So what of the results themselves, and what did they tell Beddoe, and us, about the origins of the people of the Isles? Taking his measurements of hair colour and applying the formula of the Index of Nigrescence across the whole of Britain and Ireland, the values range from 0 to 80. There is a very clear difference between the far east of Britain, by which I mean East Anglia and Lincolnshire, where the Index is lowest, and Ireland and Cornwall in the west, where it reaches its highest value, as it also does in the west of Scotland. The low values for East Anglia are also continued across Yorkshire and Cumbria and again in the far north of Scotland and the Hebrides. In all these regions the Index is about zero, which means, in practice, that there are as many blonds and redheads as there are brunettes. In other parts of England, and in Wales, the values for the Index are intermediate between the fairer east and the darker west.

The Index is a measure of hair colour alone. When it comes to eye colour, the east–west gradient is reversed. Brown eyes are commonest in the east and south, where they exceed 40 per cent in East Anglia, but also in Cornwall. In Ireland, but also in Yorkshire and Cumbria, the same counties where red/fair hair were at their highest proportion, the number of people with blue or grey eyes rises to 75 per cent. In the far north of Scotland and the Hebrides, where fair hair was common, blue or grey eyes are even commoner than they are in Ireland. When hair and eye colour are combined to produce two basic types – what Beddoe calls 'Mixed Blond' and 'Mixed Dark' – with fair/red hair and blue/grey eyes or dark eyes and dark hair respectively – the patterns reflect the individual components to some extent. 'Mixed Blonds' outnumber their opposites in the north of Scotland, east Yorkshire and Lincolnshire, while 'Mixed Darks' predominate in Wales, Cornwall and the Scottish Highlands, along with

Wiltshire and Dorset. In Ireland, especially in the west, the Mixed Blond outnumber the Mixed Dark, leading to the conclusion, when coupled with the high Index of Nigrescence, that there must be a high proportion of dark-haired people with blue or grey eyes.

Beddoe begins his conclusions in the far north. 'The Shetlanders', he says, 'are unquestionably in the main of Norwegian descent, but include other race elements also.' He draws the same conclusion for the inhabitants of Orkney and of Caithness in the far north of the Scottish mainland. He assumes the Norwegian element came with the Vikings. He also makes the strange observation that 'The excessive use of tea, the one luxury of Shetland, probably only aggravates a constitutional tendency to nervous disorders which is more prevalent among the few dark than the many fair Shetlanders.' This is the point to tell you that Beddoe describes himself as a young man 'of fair complexion, with rather bright brown wavy hair, a yellow beard and blue eyes'. Clearly it was perfectly safe for him to drink the tea.

As he works his way down Britain and across to Ireland, his observations combine preconception with perception in an extraordinarily personal record of his encounters. On Lewis, in the Western Isles, he observes the 'large, fair and comely Norse race, said to exist pure in the district of Ness at the north end of the island' and the 'short, thick-set, snub-nosed, dark-haired, often dark-eyed race, probably aboriginal and possibly Finnish whose centre seems to be in Barvas'. Barvas is 12 miles north-west of the principal town, Stornoway.

Beddoe is acutely aware of influences on the objectivity of his observations. In recording the Highlanders he says at once that 'Most travellers, on entering the habitat of a race strange to them quickly form for themselves from the first person observed some notion of the prevailing physical type.' Also he is aware that the longer he spends in a particular place, the more he can distinguish the differences between people: 'I confess that the longer I have known the Scottish Highlanders the more diversity I have seen among them'.

He also notices the presence of what he calls 'a decidedly Iberian physiognomy, which makes one think ... that the Picts were in part at least of that stock'. We will return to that particular element of the British mix later in the book. As he crosses the border into England he finds pockets of 'a very blond race in Upper Teesdale' and, further south, 'the small, round-faced dark-haired men with almond-shaped eyes ... in the vale of the Derwent and the level lands south of York', which he ascribes to either an Iberian or Romano-British origin. There is a growing feeling, as Beddoe moves around the country, that he is forming the view that dark-eyed and dark-haired people are the remnants of the indigenous Britons that were later supplemented, or displaced, by the Saxons and the Vikings. Even as he travels to the West Country, the connection is there: 'In the district about Dartmouth, where the Celtic language lingered for centuries, the Index of Nigrescence is at its maximum'. Onward to Cornwall, where 'The Cornish are generally dark in hair and often in eye: they are decidedly the darkest people in England'.

When Beddoe moves into Wales, he finds in the central region 'a prevalence of dark eyes beyond which I have met with in any other part of Britain'. The Snowdonians too are 'a very dark race', while around the coast eyes and hair are lighter. Across the Irish Sea, he records that 'the frequency of light eyes and of dark hair, the two often combined, is the leading characteristic'.

So much for the observations. What about the conclusions?

There can be a tendency among collectors to leave interpretation of their results to others, mainly, I think, for fear of being proved wrong and thus undermining their whole legacy. This is an increasing trend, but even Beddoe was shy of absolute conclusions. None the less he ventured an explanation for the fair-haired people of England, suggesting that 'the greater part of the blond population of modern Britain ... derive their ancestry from the Anglo-Saxons and Scandinavians ... and that in the greater part of England it amounts to something like half'. So there we have it. Beddoe explains the different colouring by a very substantial settlement from Saxon, Dane and Viking. The particularly light colouring in parts of Yorkshire, which we noted previously, he attributes to the impact of the Norman Conquest. However, Normans, as we will later discover, are really no more than recycled Vikings. On Ireland and the Gaelic west generally, Beddoe thought the people to be a blend of Iberians with 'a harsh-featured, red-haired race'. The Celtic 'type', with dark hair and light eyes, he ventures to suggest, may only be an adaptation to the 'moist climate and cloudy skies' which they endure.

Beddoe concludes the account of his lifetime's work with this paragraph:

> But a truce with speculation! It has been the writer's aim rather to lay a sure foundation whereon genius may ultimately build. If these remaining questions are worthy and capable of solution, they will be solved only by much patient labour and by the co-operation of anthropologists with antiquarians and philologists; so that so much of the blurred and defaced prehistoric inscription as is left in shadow by one light may be brought into prominence and illumination by another.

It is as if John Beddoe, criss-crossing the country with card and pencil in hand, calipers and tape in his knapsack, had already anticipated the arrival of genetics. How he would have loved to be alive now.

Beddoe and his contemporaries were the first to substitute observation for deduction and prejudice in exploring the origins of the people of the Isles. But, as he himself freely admits, there was still a strong subjective element in his observations of appearance. After all, our obsession with looks is ample proof of its emotional influence. It must have been almost impossible for Beddoe not to have nurtured some preconceived ideas, which, with the best will in the world, will have influenced his conclusions.

The next stage in the scientific dissection of our origins removed this subjective element completely. It began a long way from England, just as John Beddoe was enjoying a comfortable old age and the flood of honours which acknowledged the fruits of his lifetime's passion. While he posed for the frontispiece of *The Races of Britain* on the doorstep of his comfortable mansion in the early years of the last century, a scientist in Vienna was mixing the blood of dogs.

5

THE BLOOD BANKERS

If you have ever been a blood donor, or ever needed a transfusion, then you will know your blood group. You will know whether you belong to Group A, B, O or even AB. The reason for testing is to avoid a possibly fatal reaction if you were to be transfused with unmatched blood. You cannot tell, just by looking, what blood group a person belongs to. Unlike hair and eye colour or the shape of heads, blood groups are an invisible signal of genetic difference which can be discovered only by carrying out a specific test.

Though the first blood transfusions were performed in Italy in 1628, so many people died that the procedure was banned. As a desperate measure to save women who were haemorrhaging after childbirth, there was a revival of transfusion in the mid-nineteenth century. Though some patients had no problems accepting a transfusion, a great many patients died from their reaction to the transfused blood. What caused the reaction was a mystery.

The puzzle was eventually solved in 1900 by the Austrian physiologist Karl Landsteiner. After experimenting with mixing the blood of his laboratory dogs and observing their cross-reactions, he began his work on humans. He mixed the blood of several different individuals together and noticed that sometimes when he did this the red blood cells stuck together in a clump. This did not happen every time, but only with certain combinations of individuals. If this red-cell clumping was occuring in transfused patients, the blood would virtually solidify, which would explain the fatal reaction. It also explained why some patients tolerated a transfusion and showed no signs at all of a reaction.

Landsteiner interpreted the results of his mixing experiments by suggesting that people belonged to one of the three blood groups, A, B or O. Two years later a fourth group, AB, was discovered. This also explained the erratic pattern of transfusion complications. Giving a group A patient a transfusion of blood from a group A donor was fine; tranfuse a group A patient with blood from a group B donor and there would be trouble. But so long as the donor and patient blood groups were the same there was no problem.

It took a few years to discover the chemical basis for the different types of blood. The blood groups are the result of a simple genetic difference that occurs on the surface of red blood cells, the cells that carry oxygen and give blood its colour. On the outside of each red blood cell sits a molecule that can occur in two very slightly different forms, A or B. People in group A have, unsurprisingly, version A on the surface of their red cells while in group B, this is replaced by version B. In the rare AB group the cells have both A and B versions on their outer surface. People in group O have neither A nor B versions of the molecule. Their red cells are, in a sense, bald.

But these slight differences, which don't affect the efficiency or the working of the cells at all, are not on their own sufficient to cause trouble on transfusion. The problem arises because, after a few months outside the womb, the blood serum begins to build up antibodies to the *opposite* version of the molecule on their own cells. People in group A build up anti-B antibodies in their serum. Again, this does not interfere with normal everyday life. People never make antibodies to their own blood cells, so people in group A don't make anti-A antibodies, only anti-B. Since people with blood group AB have both versions on their red cells, they make neither anti-A nor anti-B antibodies while, for the same reason, people in group O, whose cells have neither A nor B, are free to make both anti-A and anti-B antibodies and they do.

The potentially fatal coagulation reaction occurs when the molecule meets its antibody. They stick to each other like glue and, what is worse, bind all the red cells into a sticky clump, the cause of all the trouble in mismatched transfusions. That's why no one makes antibodies against their own cells. They would coagulate their own red blood cells and die.

Under normal circumstances blood cells never encounter their own antibodies, but transfusion opens up that possibility. Transfuse a group A patient with blood from a group B donor and the antibodies will play havoc. Two things happen. The group B cells from the donor are coagulated by the anti-B in the patient's serum and the anti-A in the donor's serum clumps the patient's own cells. Group O blood is really bad news because its serum contains both anti-A *and* anti-B which will attach the cells of any other blood group. However, as good methods were developed to separate the donor's cells from the liquid serum, things got a bit easier. Group O cells, separated then rinsed free of antibody-containing serum, can be transfused into any patient, and if red cells are all you need that's fine. Group O is the universal red-cell donor, as long as you wash them thoroughly first to remove the serum antibodies. If you need serum, not cells, then a transfusion of AB serum, which is free of antibodies, will suit any patient whatever their blood group.

Once all this was understood, it was easy to see why so many transfusions failed. Without knowing in advance the blood group of donor and patient, blood transfusion was a really hit-and-miss affair. At least that was the case in Europe. Stories that the Incas of Peru had been successfully performing a form of blood trans-fusion without any adverse reactions were initially dismissed as nonsense. However, when it was discovered that practically all native South Americans were in blood group O, it no longer sounded so incredible. If Inca donor and Inca recipient were both in group O, as most were, then trouble-free transfusions are exactly what would be expected.

Before the First World War, blood transfusions were a personal business. Willing friends and relatives of the patient would be tested to find someone in a compatible blood group. The donor would then come to a hospital, usually the operating theatre if surgery was the reason for the transfusion, and be bled right next to the patient, who then immediately received the fresh blood. The huge increase of blood transfusions needed to treat the battlefield casualties of the First World War led directly to the setting up of blood banks and the recruitment of donors along modern lines. Under the right conditions it was found that blood could be stored for several days without losing condition and there was no need to transfuse casualties immediately with absolutely fresh blood.

Volunteer donors were bled at remote sites and the blood was despatched to the field hospitals at the Front to be matched and used as required. This soon became a large-scale activity and with it came the necessity for accurate records. Each army had its blood bank and they soon began to accumulate blood-group records from very large numbers of soldiers, each of whom was routinely tested in anticipation of being called either to give blood as a donor or to receive it as a casualty.

Hanka Herschfeld, from the Royal Serbian Army, was the medical officer in charge of the Allied blood bank on the Balkan Front. Her husband, Ludwig, had been one of the scientists who, before the war, had helped to work out the way the different blood groups were inherited. With this background it is no surprise that they became curious about the accumulating results from the blood bank. The Allies drew their troops from all over the world and the Herschfelds noticed that the frequencies of the blood groups in the soldiers of different nationalities were often quite different from one another. Certainly they were still all either A, B, AB or O, but the proportions of each were different depending on where they were from. For example, far more Indian Army soldiers belonged to blood group B than did Europeans, who were, symmetrically, higher in the proportion of group A.

The Herschfelds interpreted these differences in blood-group frequencies as having something to do with the distant origins of these different nationalities – and they were right. But in their now famous paper published in the leading medical journal the *Lancet*, just after the war, they went too far and divided the world into two separate races. Race A came from northern Europe, while Race B began in India. The varying blood-group proportions seen in the soldiers of different nationalities were explained by the mixing as people flowed outwards from these 'cradles of humanity', as the Herschfelds called them, to populate the world.

A rare display of
English national
identity. England
supporters at the
2006 World Cup
in Germany.

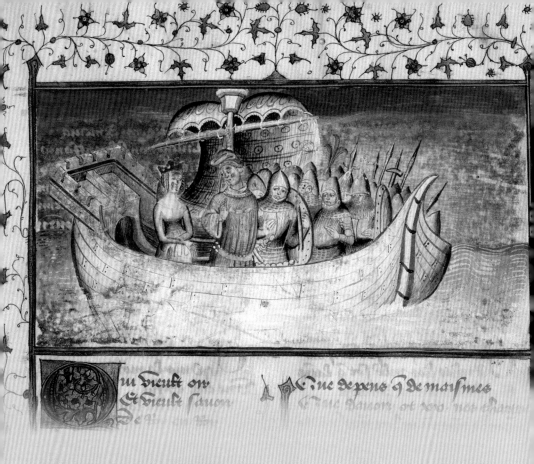

Du vient on
Et vient sauo...

A P...ue depens q de mo...fmes

Above: Brutus the Trojan sets sail to discover Britain and become its first king – according to Geoffrey of Monmouth, at any rate.

Left: The coronation throne at Westminster Abbey with *Lia Fail*, the Stone of Destiny, before its return to Scotland.

OPPOSITE PAGE: (clockwise from top left)

Tintagel Castle on the Cornish cliffs, where Uther Pendragon deceitfully seduced Eigr and King Arthur was conceived.

Merlin before King Vortigern as he prophesies the subterranean lake containing the White and Red dragons.

King Arthur, seriously wounded at the battle of Camlan, well on the way to recovery thanks to Morgan Le Fay and her attendants on the Isle of Avalon. Or is he?

Glastonbury Abbey, scene of the ritual reburial of King Arthur's remains by Edward I on Easter Day, 1278.

Right: Mask made 9,500 years ago from the skull and antlers of a red deer, recovered from the Mesolithic site at Starr Carr, East Yorkshire.

Above: The defensive ditches of the Iron Age fort of Maiden Castle, Dorset, which fell to the II Augusta legion under Vespasian following the Roman invasion of Britain in AD43.

Below: Stonehenge, iconic religious centre of the ancient Britons. The massive stones were transported 250 miles from the Preseli mountains of West Wales, about 4,200 years ago.

Exquisite gold and garnet cloisonné buckle from a sword belt recovered from the Saxon ship burial at Sutton Hoo, Suffolk. Made in the early seventh century but still in mint condition.

Aerial view of the Roman fort at Richborough, Kent, where the main invasion force landed in AD43. The platform with the cross marks the foundations of the 26-metre-high triumphal arch faced in white Carrara marble, through which all official visitors entered the province of Britannia before heading inland along Watling Street.

Saxon ceremonial helmet inlaid with bronze gilt belonging to Raedwald, King of East Anglia, recovered from the Sutton Hoo burial.

The unopposed landings by the Normans at Pevensey in 1066, as depicted in the Bayeux tapestry. King Harold was busy fighting a Norse army near York at the time and had to dash south to fight the Normans under Duke William at the Battle of Hastings.

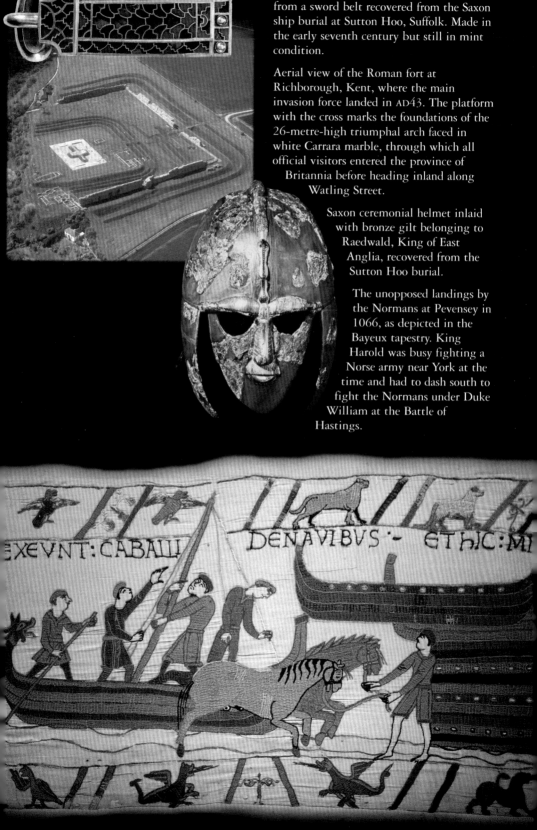

Full moon over the standing stones of Callanish
on the Isle of Lewis in the Outer Hebrides, one
of the breathtaking megaliths constructed along
the Atlantic fringe from Iberia to the Shetland
Isles about 5,000 years ago.

OPPOSITE PAGE: (clockwise top left to bottom)

The exposed interior of one of the 5,000-year-
old houses at Skara Brae, Orkney, complete
with domestic furniture, hearth and even
fishtanks to keep the lobsters fresh!

The central chamber of the passage grave at
Maes Howe, Orkney, showing the vaulted roof
built 5,000 years ago with the conveniently
flat sandstone slabs from the locality.

Hadrian's Wall, Northumberland, built in
AD122 to mark the final frontier of the Roman
Empire – and to keep the belligerent Picts
out.

The settlement at Jarlshof, Shetland, where
the rectangular Viking longhouses have
incorporated the earlier Pictish roundhouses.

The eighth-century Pictish symbol stone at
Aberlemno, near Perth, depicting a battle
between the Pictish King Bruide and an
invading army under Ecgfrith, King of
Northumbria. The Picts won. The hole in the
stone was drilled to make it easier to move!

Mousa Broch in Shetland, the largest of the
characteristic Pictish fortified dwellings built
to a standard design between 2,100 and 1,900
years ago.

Right: Robert Knox MD, anatomist, surgeon, author – and venomous racist.

Centre: 'A Celtic groupe; such as may be seen at any time in Marylebone, London.' An engraving from Robert Knox's *The Races of Men*, in which he makes his attitude to the Celts abundantly clear.

Below: John Beddoe towards the end of a long life spent recording the physical features of the people in every corner of the Isles.

John Beddoe

A page from Beddoe's album of photographs showing an ironworker from York of Welsh parentage coupled with a short handwritten description and a serial number. Beddoe was an avid and systematic collector.

Griffith Llewellyn at 50 dark eyes. hair nearly black. both ear lobes welded. an ironworker. both parents natives of Pembrokeshire, S. Wales.

Taken in York Castle August 82. J. Anderson.

Their *Lancet* paper is a classic, and rightly so. It was the first of its kind and it opened up an entirely new field of research in anthropology. It follows on from the implicit assumption in John Beddoe's research on physical appearance that inherited features can be used to explore the origins of people. Compared to the work on hair and eye colour, skull shape and so on, blood groups come one step closer to the fundamental controller of genetic inheritance, DNA. However, no one knew about the way DNA conducted the business of inheritance at the time the Herschfelds were at their peak, nor for several decades afterwards. Blood groups, though still an indirect manifestation of the underlying DNA, were a definite improvement on the earlier, subjective parameters which were all that were available to John Beddoe and his Victorian contemporaries.

For one thing, it completely removed prejudice and human error from the equation. Blood groups are tightly defined and there is no overlap between them. No matter who does the tests, someone in group B will always be in group B. It doesn't alter with age. There is no room for doubt, at least not about the accuracy of the observation. But there is also a noticeable shift in the tone of the reports. There are no longer any barely concealed inferences of racial character, like the free-spirited, fair-haired Saxon who will not be tied to the drudgery of an urban existence but would rather make his fortune overseas, or the morose, dark-haired Shetlander driven to despair by drinking too many cups of tea. All that nonsense vanishes, as it is very hard to get worked up about the comparative personal characteristics of one blood group over another. The American physician William Boyd, who extended the Herschfelds' work around the globe, expressed this new sense when he wrote, 'In certain parts of the world an individual will be considered inferior if he has, for instance, a dark skin but in no part of the world does possession of a blood group A gene exclude him from the best society'. As a group A myself, that comes as something of a relief.

The Herschfelds' final legacy was less glorious. Their grand conclusions about the dual origins of humanity turned out to be completely wrong. It took the discovery of other blood group systems, unimportant in transfusion, and the amalgamation of results from several of them to get a more reasonable interpretation of human evolution. Slowly the searchlight illuminating the 'cradle of humanity' turned away from Europe and Asia and settled firmly on the plains of East Africa.

After the Second World War, the task of sifting through the, by then, thousands of sets of blood-group data from transfusion centres all over the world settled on the shoulders of one man, Arthur Mourant. Born in Jersey in the Channel Islands, and emotionally attached to it throughout his life, Mourant wanted at first to become a psychoanalyst and, in order to do so, he had to enrol as a medical student. He was forced to abandon his ambition after he underwent analysis himself and was judged to be 'emotionally unsuitable' to continue with his training. Despite this disappointment, he did not abandon his medical training. During the Second World War his medical school, St Bartholomew's in the City of London, was evacuated to Cambridge and he found himself working with one of the greatest geneticists of the twentieth century, R. A. Fisher.

Fisher, among many other interests, was engaged in a bitter rivalry with the American physician Alexander Wiener on the inheritance of the Rhesus blood groups. Rhesus is another type of human blood group, discovered, as its name implies, through research in Rhesus monkeys. Unlike Landsteiner's comparatively simple ABO system, the genetics of the Rhesus blood group are fiendishly complicated. There was a furious race to unravel the genetics by following the Rhesus group through families, and the young Mourant was assigned to the task of finding and typing suitably large families. He quickly found the solution to the inheritance, thanks to his particularly fruitful research on a local East Anglian family. Fisher was understandably elated by this triumph and Mourant soon found himself an established member of Fisher's team with a bright career ahead of him. This career he dedicated to extending William Boyd's pre-war surveys on the geographical distribution of the ABO blood groups around the world. He also set out to enlarge on Boyd's work by including as many of the newly discovered blood groups, like Rhesus, as possible and to build up the most complete maps of blood-group distributions in every part of the globe. Mourant, like Beddoe before him, was an extremely avid collector of information.

In front of me as I write is his final masterpiece, *The Distribution of the Human Blood Groups*, published in 1976. It is an impressive tome, 6 centimetres thick and weighing 2.4 kilograms. It has 1,055 pages, 3,179 bibliographic references, 661 pages of tables and several pages of beautifully drawn maps. It has the gravitas of a life's work. To give you a taste of the scope of Mourant's encyclopaedic enterprise, there are tables of Rhesus blood-group data from the Bilwa, Sanpuka and Ulwas tribes of Nicaragua, the ABO blood groups of 13,000 blood donors from Benghazi in Libya, the MNS (another blood group) results from French and Spanish Basques and the Duffy (yet another) blood-group results from hundreds of New Guinea highlanders.

But what did it all mean? Surely there was enough information here to resolve any lingering uncertainties about the whereabouts of the 'cradle of humanity' and how our ancestors had moved from there to populate the planet. But even a cursory inspection of the maps shows the optimism is misplaced. Working along the commonsense lines that if two peoples have similar proportions of the different blood groups, then they are more likely to be related than two with very different proportions, you soon run into trouble. For example, the highest frequencies of blood group A are found in two very different parts of the world: among native Australians on the one hand and Saami reindeer-herders of northern Norway on the other. It would be preposterous to propose that these two peoples, about as far away from each other as it is possible to be, were closely related and shared a recent common ancestry. However, when you factor in the results of the other blood groups, the relationship becomes far more reasonable. The native Australians might have the same proportions of the ABO blood groups, but the composition of the other groups, like Rhesus, MNS and Duffy, are utterly different. Bit by bit, blood groups began to draw out connections between the different peoples of the world, including western Europeans and the people of the Isles.

The basic pattern which Mourant found across western Europe, the area of most relevance for *Blood of the Isles*, showed that across the whole region west of the River Elbe in Germany, group A is high and group B comparatively low. East of the Elbe the opposite is the case. There is a gradual shift from B to A, a so-called genetic cline, as we get closer to the Isles. In the Isles themselves Mourant was able to call on absolutely vast amounts of material, both from his own unit, by now incorporated as an official laboratory of the Medical Research Council, the UK's government funding agency for medical research, and from other published works. Among these were detailed records of blood-group frequencies from Ireland, Wales and the Scottish Highlands.

Mourant gave the task of collecting all the records from the 'missing' bits – that is to say England, lowland Scotland and Northern Ireland – to his long-term assistant Ada Kopec, who, with a librarian and a secretary, made up the entire staff of the Blood Group Centre. It is plain from reading the account of this mammoth piece of assimilation and statistical comparison of a grand total of 477,806 results that Ada Kopec was far more concerned with mathematical manipulation of the figures than with explanation. Indeed it is left to Mourant himself, writing in the Foreword, almost to excuse his assistant from any genetic or anthropological interpretation, which, he writes, 'will have to be made by others'. Fortunately, there were 'others' prepared to stick their necks out.

The conclusions of Mourant and Kopec's gigantic enterprise can be summarized very concisely. In Ireland there are very high levels of blood group O, the highest in Europe. The further west you go, the higher the group O proportions. And, as elsewhere in Europe, where O is high, A is low and vice versa, so in the eastern counties of Ireland, where O is lower than in the west, A is higher. The differences in different parts of Ireland are not dramatic, but because the number of individuals taking part is so high, the figures can be relied upon to be statistically reliable. So whereas in County Clare, in the far west of Ireland, 80 per cent of people are in group O, this drops to 73 per cent in County Wexford in the south-east, with a mirror-image result for blood group A. Turning to blood group B, there is a slight reversal of the trend across the rest of north-west Europe. Instead of following the rule of the further west, the lower the proportion of B, there is a distinct and statistically significant rise in the far west of Ireland compared to the east. It goes from 6.6 per cent in Wexford to 8 per cent in Kerry.

Now comes the explanation. According to Professor Geoffrey Dawson from Trinity College Dublin, the high levels of A in south-east Ireland are a direct result of successive waves of immigration into Ireland from England. In the first of many papers on the blood groups of Ireland, written in the 1950s, Dawson kicks off with a summary history of Ireland. He explains the blood-group changes from east to west by the Anglo-Norman invasion in the twelfth century, which we will revisit in a later chapter, and by attempts to settle English immigrants under Queens Mary and Elizabeth in the late 1500s.

I much prefer it when authors do advance a theory to explain their results, rather than leave it to others. But how can Dawson possibly know this is the reason for the blood-group differences? Why could they not be equally well explained by other movements of people in prehistoric times? Or by a mixture of both? The whole thrust of the explanation is based on historical events that we already know about. If we had not had a reasonable explanation to hand, would the blood-group evidence be strong enough to come up with one on its own? I really doubt that. Instead of proposing something completely original, the genetic data is rationalized and fitted in to what we already suspect from other sources.

The rationalizations reach their peak in relation to Iceland. Iceland was unoccupied until the late 800s when the systematic settlement from Scandinavia began. The language, the culture, even the written histories recorded in the Icelandic sagas, including the *Histories of Settlement*, leave no one in any doubt that the great majority of settlers were Norse. And yet, the blood-group proportions in Iceland are very different from those of modern-day Norway and almost identical to those of Ireland, as the table shows.

	A	B	O
Iceland	19	7	74
Norway	31	6	62
Ireland	18	7	75

By any token, the only conclusion from the blood-group composition is that Iceland was not settled from Norway at all. Far more likely, from the blood-group results, is a wholesale settlement from Ireland or somewhere else with similar blood-group proportions, like parts of Scotland. As we will see in a later chapter, there is at least a partial explanation for this discrepancy, but that is not the main message I want to get across here.

Faced with this disagreement in the blood results, instead of having the confidence to overturn the theory of Norse settlement, Mourant tries to rationalize by finding Scandinavian 'homelands' that might heal the discrepancy. He cites parts of western Norway around Trondelag that have a blood-group composition a little more like Iceland than the rest of the country, then reports an isolated population in northern Sweden in the province of Vasterbotten with an even more Icelandic composition. Northern Sweden isn't even close to the Atlantic and no traditions link it to the settlement of Iceland. Mourant then highlights an old settlement at Settesdal in southern Norway with 'Icelandic' blood-group compositions. Finally, to resolve this awkward disagreement, he suggests that the modern-day Scandinavians are the descendants of people moving in from the south and east who displaced the Vikings and drove them to settle in Iceland.

All of these attempts to resolve the disparity between, on the one hand, mountains of cultural and historical evidence on the Scandinavian origin of the Icelanders, and the blood-group results on the other, highlight a fundamental weakness in the value of using blood groups to infer origins. If the results from the labs agree with what you already believe about the origins or make-up of people, then there is a cosy feeling that the genetics, archaeology and history are all in agreement with each other. But when they do not there is a temptation to fabricate an agreement with increasingly unlikely scenarios, as with Iceland.

I suspect the same has been done in the south-west corner of Wales. The southern part of Pembrokeshire surrounding the deep-water inlet of Milford Haven delights in the sobriquet of 'Little England beyond Wales', a reference to the anglicized place-names and the long use of the English as opposed to the Welsh language. The levels of group A in this small region of Wales are 5–10 per cent higher than in the surrounding areas. It is known that Henry I forcibly transferred a colony of Flemish refugees fleeing political repression in Holland and Belgium to the area in the early twelfth century. The high levels of blood group A have been attributed to this historical influx and are often quoted in popular accounts as a classic success of blood grouping confirming history. This is despite the levels of blood group A in the Low Countries not being particularly high. However, a very different explanation was favoured by the Welsh scientist Morgan Watkin, the man who originally noticed the high proportion of group A in parts of Pembrokeshire. He put it down to a substantial Viking settlement in the region, despite the fact that there is very little in the way of archaeology or place-names to support it. But the fact remains that, even after thousands of blood samples from Wales and hundreds of thousands from all over Britain and Ireland, it is still impossible to decide whether the unusual blood-group composition of this part of Wales was caused by rampaging Vikings or by a few cartloads of Belgians.

The root of the problem is that, despite there being vast amounts of very reliable data, blood groups just do not have the power to distinguish these two theories, nor the power to propose new ones that might fly in the face of historical or archaeological evidence. Blood groups, despite the advantage of objectivity, are a very blunt instrument indeed with which to dissect the genetic history of a relatively small region like the Isles. Fortunately, we can sharpen our genetic scalpel. Now we can do something that William Boyd, Arthur Mourant and the others could not. We can move to the next stage and take the last step towards the final arbiter of inheritance. We can move to the DNA itself.

6

THE SILENT MESSENGERS

Whatever their shortcomings as a guide to the past, the fact that blood groups are 100 per cent genetic makes it self-evident that they are inherited from ancestors. They are not DNA, but they are the expression of DNA. You may like to compare the relationship between DNA and blood groups like this. When you listen to a piece of music you are not hearing the written notes themselves, but the expression of the notes as interpreted by the musicians. Our inherited features, both those we notice, like hair and eye colour, and those, like blood groups, that we need tests to reveal, are the music we hear. The DNA is the equivalent of the notes on the sheet, which the musicians are reading to produce the music.

Arthur Mourant and his fellow blood-groupers were too early to see the sheet music on which the blood-group notes were written, but they knew from the way it was inherited in families that it must be very simple. The four different blood groups A, B,

AB and O are the expression of three versions of a single gene, a single piece of DNA. Once it became possible to read the notes behind the music, the true cause of the blood groups was revealed to be very slight changes in the DNA of the blood-group gene itself. DNA is a coded message in the form of a sequence of four slightly different chemicals attached to each other. If you think of it as a very long string of beads, where each bead is one of these DNA chemicals, then that will give you an idea of what a strand of DNA looks like. Now imagine that there are four different colours of bead on the string, each one representing one of the four DNA chemical bases, as they are called. You can see how the string of beads might become a code purely by virtue of the sequence in which the different coloured beads are arranged. The DNA of the blood-group gene is about 1,000 beads, or bases, long.

Though it calls the shots, DNA doesn't actually do the work in the body, just as the notes on a sheet of music need musicians to be heard. DNA is the code that tells cells, all of which contain DNA, what to do. Just as notes on a musical score tell the orchestra what to play, DNA tells cells which proteins to make. And it is proteins that build and run the body. Proteins are made up of amino-acids arranged in a specific linear sequence and it is this sequence of amino-acids that gives the protein its particular properties. No two proteins are the same. The protein collagen, for example, has a very strong and rigid structure which it needs to do its job in strengthening bones and teeth. That strength is a direct result of the way the amino-acids are arranged, just as the oxygen-carrying capacity of haemoglobin comes about by the particular sequence of its own amino-acids. The same goes for the blood-group protein that sits in the membrane of red blood cells. It is all down to the sequence of amino-acids.

The DNA instructs the cell how to make proteins through the coded instructions held in the sequence of the coloured beads on the string. Cells know how to interpret this code and how to translate the DNA sequence into the amino-acid sequence of a protein. The differences between alternative versions of the same gene, which are what produce the three different blood groups, are caused by mutations. This is when, very rarely, there is an error in copying the DNA. A bead suddenly changes colour and the DNA sequence changes slightly. Cells read the new sequence like the mindless automata they are. They don't realize that they are now producing a slightly different version of the protein, which may have different properties. They just do as they are told.

Most mutations happen when DNA is being copied. Since every cell contains a full set of DNA, it has to be copied every time a cell divides. We all start off as a single cell, a fertilized egg, and grow from that by cell division to an adult with 10 million billion cells, so there is an enormous amount of DNA copying going on and plenty of opportunity for DNA mutation. However, the fidelity of copying DNA is absolutely fantastic, and of course it needs to be. If it were as poor as the average photocopier, by the time the fertilized egg had divided and divided to produce at first an embryo, then a foetus, then a baby, the DNA instructions would become so fuzzy that every child would be born with every genetic disease under the sun – if he or she ever got born at all. To prevent this happening, there are proofreading and editing mechanisms which scan the newly copied DNA to make sure it matches the original sequence. All of this is to reduce the chance of mutation. And in this we are very successful. On average, a DNA base mutates only once in every thousand million times it is copied. Even so, this minuscule error rate is enough to produce all the genetic variation in our own species and in every other living creature that we see in the world around us. Mutation is the life-blood of evolution.

Without mutation, there simply is no evolution. Most of the time mutation, even when it occurs, has absolutely no effect. Very occasionally, though, mutations do drastically affect the working of whatever protein the gene is in charge of – and that is how devastating inherited diseases can begin their life. In my earlier career as a medical geneticist, working as I did with inherited bone diseases, I saw many patients whose bones would fracture at the slightest knock. They were badly deformed and often unable to walk – but often astonishingly cheerful and optimistic. Their disease, called osteogenesis imperfecta, a very serious form of brittle-bone disease, was caused by one of these random mutations in a bone collagen gene. But instead of making a harmless change to the DNA sequence, in these patients the mutation had hit a crucial DNA base in the collagen gene. The mutations in these patients, even though they change just a single DNA base, completely alter the structure of the collagen, turning it from an extremely strong protein into the biological equivalent of putty.

Mutations can be good, bad or indifferent. Most are indifferent, like the mutations which produce the different blood groups. A few are bad, as in the brittle-bone patients. Vanishingly few are good, in the sense that they improve the way the protein works. On the whole the bad mutations are eliminated pretty swiftly as people with inherited diseases die or have fewer children. Good mutations can find themselves increasing from one generation to the next if they aid the survival of the people that carry them or help them have more children. Indifferent mutations, and they are in the majority, have no influence one way or the other on survival or success in breeding. They just get passed from one generation to the next, their fate entirely out of their hands. They risk elimination if they end up in someone who has no children or can do well if they find themselves in a large family. They might lead less dramatic lives than the mutations that bring success or devastation. But it is these, the silent passengers of evolution, that are its most articulate chroniclers. This is precisely because they cause no ripples, they are unseen by natural selection and are neither promoted nor destroyed by its attentions. But nowadays, thanks to the breakthroughs of the last twenty years, we can see them in the read-out from the DNA analyser. And we can use them to trace our ancestry.

While Arthur Mourant did what he could with the very limited number of blood groups, there is almost no limit to the amount of different DNA sequences that we are now able to detect. It is this massive increase in our ability to distinguish one person's DNA from another which has made all the difference in our ability to trace our ancestry and discover our genetic origins. But with all this choice, which were going to be the best genes to concentrate on, and why?

During my work on ancient bones I wanted to give myself the best chance of recovering DNA so I chose to focus on a rather unusual piece of DNA. Most of our DNA is contained within the cell nucleus, attached to tiny thread-like structures called chromosomes. This is where the collagen genes, the haemoglobin genes and the blood-group genes reside. For all of them, as for most of our 'nuclear' genes, we have only two copies in each cell, one from each of our parents. However, outside the cell nucleus, though still inside the cell membrane, there is a different source of DNA altogether. In the liquid cytoplasm surrounding the nucleus are tiny particles called mitochondria. These particles control many of the steps in aerobic metabolism and they have an interesting evolutionary history, having once been free-living bacteria. From our point of view at the time, where this DNA had come from and what it did was unimportant. What counted was that there was far more of it in the average cell, maybe a thousand times more, than the DNA of any of the nuclear genes. If only a few cells survived in the ancient bones, targeting mitochondrial DNA would maximize our chances. It turned out to be the right decision, and we found mitochondrial DNA in the first batch of bones we tried. It is still extremely hard to recover nuclear genes from ancient specimens, while getting out the mitochondrial DNA is now almost routine.

As well as its abundance in each cell, mitochondrial DNA (or mDNA for short) has two other outstanding properties to recommend it as a window into the human past. Firstly, it mutates about twenty times faster than regular nuclear DNA. The error-checking mechanisms in mDNA are much less vigilant than they are in the nucleus. Our species has been around for about 150,000 years and, although this seems to us like a very long time, the nuclear DNA mutation rate is so low that the vast majority of it is completely unchanged since that time. In a typical stretch of nuclear DNA 1,000 bases long, nineteen out of twenty people will have exactly the same sequence. Within the same sized stretch of mDNA, almost everyone is different.

The second excellent feature of mDNA is its very unusual inheritance pattern. As we have seen, most of the nuclear genes are inherited equally from both parents. You have received one copy of each nuclear gene from your mother's egg and one from your father's fertilizing sperm. But you got *all* of your mDNA only from your mother, and for one very simple reason. Compared to sperm, eggs are huge cells, bulging with cytoplasm, which is crammed with a quarter of a million mitochondria. Sperm do have a few mitochondria, about a hundred, in what is called the mid-piece, which connects the sperm head, containing all the nuclear DNA, to the tail. The thrashing tail needs the aerobic energy output of the mitochondria in the mid-piece to fuel its progress towards the egg.

But once the successful sperm penetrates the egg to deliver its precious load of nuclear DNA, its mitochondria are not only vastly outnumbered but are deliberately destroyed. This is why, although the fertilized egg contains nuclear DNA from both father and mother, all the mitochondria, and so all the mito-chondrial DNA, is from the mother.

The process is repeated generation after generation after gen-eration. Nuclear DNA comes from the father and mother, mDNA only from the mother. Consider your own mDNA for a moment. It is powering your aerobic metabolism in every cell – from the cells in your retina which collect the focused image from the page, to the muscles in your arm that turn the pages, to the cells that are burning fuel to keep you warm. All these functions are controlled by your mDNA which, because of its unusual inheritance, you have got only from your mother. Who got it from her mother. Who got it from her mother and so on. At any time in the past, be it 100, 1,000, even 10,000 years ago, there was only one woman alive at the time from whom you have inherited your mDNA. Even though I have known this for years it still amazes me to think about it.

The combination of plenty of genetic variation with its matrilineal inheritance makes mDNA the perfect guide to the human past. But it needs to be complemented, because it can tell only one side of the story. Mitochondrial DNA can only tell the history of women. Very fortunately, there is a piece of DNA which can do the same for men. This companion guide to our genetic history could not be more different. This is the piece of DNA that is entirely male. It is the Y-chromosome.

Inside the nucleus of every human cell are a total of forty-six chromosomes. Forty-four out of the forty-six carry on them the great majority of the 10,000 genes that build and run our bodies. They include the blood-group, collagen and haemoglobin genes we have already met and many, many more. They direct almost everything, from aspects of our physical appearance like eye and hair colour, to our immune systems, to our innate psychological and emotional make-up. In everybody, male and female, these forty-four chromosomes come in pairs and are inherited from both parents, twenty-two from one, twenty-two from the other.

The other two chromosomes, called X and Y, are different in that they are not always inherited from both parents. And not everybody has both of them. Females have two X-chromosomes and men have one X-chromosome and one Y-chromosome. In the official notation of genetics, women are XX and men are XY. However, despite what I have come to appreciate that most people believe, the X-chromosome has nothing directly to do with sex. Women are not women because they possess two X chromosomes – the truth is far more interesting. Women are female because they don't have a Y-chromosome. How can that be?

Looked at under the microscope, the X and Y chromosomes look quite different. Both are the same shape, like tiny threads, but the X-chromosome is about five times as long. The differences between X and Y don't stop there. Thanks to the output from the Human Genome Project we now have the DNA sequence for both chromosomes. The larger X-chromosome is very like the other forty-four chromosomes. It carries about 1,000 genes which control a range of different cellular activities. The

Y-chromosome, on the other hand, is a genetic wreck with only twenty-seven genes that appear to be working properly. The rest of the chromosome is made up of long stretches of so-called 'junk' DNA. This is DNA that, unlike genes that do things, has no known function. It is just there. The evolutionary implications for this tremendous difference between X- and Y-chromosomes are fascinating, but not especially relevant here. What does matter is that just one of the twenty-seven active genes on the Y-chromosome, the sex gene, is what makes males.

For the first six weeks of life, there is no visible difference between male and female embryos. At about that time, the sex gene on the Y-chromosome switches on. This sends a signal to a whole series of other genes situated on other chromosomes, which, between them, actively divert embryonic development away from female and towards male. Embryos that don't have a Y-chromosome just carry on along the normal female development pathway and are born girls. The X-chromosome has nothing to do with it. Men truly are genetically modified women.

This mechanism for deciding sex which humans have inherited from their distant mammalian ancestors creates the second of our guides to our genetic origins. Men carry both an X- and a Y-chromosome in all of their cells – except mature sperm. Sperm occur in two different genetic forms, indistinguishable under the microscope and in their swimming capabilities. Stem cells in the male testis are dividing furiously to keep up the supply of sperm and like the other cells in the body have the XY combination of sex chromosomes. At the final division, the cell divides one last time but the resulting sperm only get one of the sex chromosomes, not both. Half the sperm receive an X-chromosome from this division while the other half get a Y-chromosome. The sex of the child entirely depends on which sort of sperm wins the race to the egg. If it's got an X-chromosome then the egg, which already has one X-chromosome, becomes XX after fertilization, develops as a female embryo and is born a girl. If, on the other hand, the winning sperm contains a Y-chromosome, the fertilized egg becomes XY and develops into a boy. The simple conclusion is this: Y-chromosomes get passed down the male line from father to son.

Looking backwards, if you are a man, you got your Y-chromosome from your father, who got it from his father. Who got it from his father. Sounds familiar? It is the mirror image of the inheritance pattern for mitochondrial DNA. The Y-chromosome is the perfect complement to mDNA, telling the history of men. But does it have enough genetic variability to be practically useful? It took a very long time to find any mutations at all on the Y-chromosome. For those scientists involved, and thankfully I wasn't one of them, it was a frustrating few years. In one of the first studies looking for diversity among human Y-chromosomes, 14,000 bases were sequenced from twelve men from widely scattered geographical localities. Only a single mutation was discovered. Another lab sequenced the same 700-base segment from the Y-chromosomes of thirty-eight different men and didn't find a single mutation in any of them. At long last, and helped by an ingenious technique for finding the elusive mutations, the Y-chromosome began to show its genetic jewels. Slowly, slowly, mutations that had changed one DNA base to another were teased out of the otherwise barren desert of uniformity.

With these two pieces of DNA we have the perfect companions for our exploration of the genetic past. One follows the female line, the other tracks the male genealogy. What could be better? They had been my guides in Polynesia and in Europe and I knew them well. Among their many qualities is that they both group people into clans. When my colleagues and I had been trying to make sense of the mDNA results from Europe in the early 1990s, we noticed that the 800 or so samples from volunteers from all over Europe fell into seven quite distinct groups based on their mDNA sequences.

Unlike the chromosomes in the cell nucleus, which are straightforward linear strings of DNA, mitochondrial DNA is formed into a circle, which is a hangover from when the mitochondria themselves were free-living bacteria. The human mitochondrial DNA circle is exactly 16,589 DNA bases in length, but fortunately it is unnecessary to read the entire sequence. Most of the mitochondrial DNA circle is taken up with genes that code for the enzymes involved in aerobic metabolism, which is the prime function of mitochondria in the cell. Because these enzymes have a very particular structure, decided by their amino-acid sequence, mutations in the genes which alter the amino-acid sequence almost always diminish or destroy the enzyme activity. The individuals who are unfortunate enough to experience these mutations in their mDNA usually die. Aerobic metabolism is such a vital part of life that we cannot tolerate even the slightest malfunction. The genetic result is that because these individuals rarely live long enough to have any children, the mutations are not passed on to future generations. If all mDNA mutations behaved like this, we would never find any genetic differences between individuals and it would be quite useless as a guide to the past because everybody's mDNA would be the same. However, fortunately for our purposes, not all mDNA does code for these vital metabolic enzymes.

Approximately 1,000 of the 16,589 DNA bases in the mDNA circle have a different function altogether, one that does not depend on the precise sequence. This stretch of DNA is called the 'control region' because it controls the way mDNA copies itself during cell division. Fortunately for us, part of this control region comprises a stretch of 400 bases whose precise sequence is un-important. It is really just a piece of genetic padding. It must be there and it must be 400 bases long for the control region to work properly, but it doesn't seem to matter what these 400 bases actually are. This is the complete opposite to the parts of mDNA that code for the metabolic enzymes, which, as we have seen, need to have a very particular sequence. The vital consequence for us of this tolerance in the DNA sequence of the control region is that when a mutation happens it doesn't affect the performance of the mitochondria at all. Instead of killing the individual who carries it, the control-region mutations just carry on unnoticed through the generations, and we can find them.

During our work in Europe it was the mDNA sequences that we found in the control region that showed us that there were seven principal groups. Within each group, everybody shared a particular set of control-region mutations. The notation that we used to describe these mutations was as simple as we could make it. We chose one particular sequence as our 'reference sequence'. If we use the metaphor of DNA as a word, then the reference sequence is its standard spelling. The sequence we chose as the standard was the one we most frequently encountered in Europe. If a particular mDNA sequence differed from the reference at the 126th base of the 400 in the control region, then it was denoted simply as 126. If there was another mutation at the 294th position, then the notation became 126, 294. We found a lot of people who shared this particular combination of mutations and they formed one of our seven groups. In other groups there were different sets of 'signature' mutations. However, within the groups like the one defined by mutations at 126 and 294, there were plenty of other mutations as well. While about a third of people within the group had just the bare minimum of 126, 294, the rest had one, two, three or even more additional mutations.

By looking for the signature mutations it was fairly easy to place any individual DNA into one of the seven groups. Occasionally we would find individuals where one of the signature mutations had changed back to the original reference, but on the whole it was quite straightforward. But what did these groups actually signify? It had to mean that everyone within the same group must be related to one another through their matrilineal ancestors, which was the line we were following with mDNA. If two people in the same group had been able to follow their maternal ancestry back in time through their mothers and their mother's mothers and so on, at some point they would converge. There would have been a woman living in the past who was the common ancestor of both of them. It then struck me, after what now feels like an embarrassingly long time, that if this worked for two people in the same clan it must, by an inevitable logic, also work for the entire clan. If one were to trace back *all* the maternal lines of everybody within each clan, they would end up with just one woman. There was no alternative. Amazing as it sounds, this has to be true.

I realized at once that these clan mothers, as I called them, were not some kind of theoretical ancestors, but real living, breathing women. No, not just women, they were mothers as well. Mothers who had survived and whose children, or at least whose daughters, had survived and who in turn had survived and had daughters and so on, right down to the present day. Though men have mDNA, they do not pass it on to their children, but they do inherit it from their mothers. Originally to emphasize to myself that these clan mothers were real individuals, I gave them names, each of which began with the letter by which the seven different groups were by then known among scientists. So the clan mother of Group H became Helena, T became Tara, J became Jasmine, X became Xenia, V became Velda, K became Katrine and U became Ursula. Over 95 per cent of native Europeans are in one of the seven maternal clans, and so it followed that these seven women were the maternal ancestors of almost all Europeans. As soon as I had given them names, they came alive and I had to know more about them. I became quite desperate to build up a picture of their lives. I wanted to know all there was to know about these seven women, the women who soon came to be known as the Seven Daughters of Eve.

The first thing I wanted to know was how long ago these seven women had lived. Were we talking about hundreds, or thousands, or tens of thousands of years ago? The answer came by looking at the extra mutations within the clan. Taking the clan defined by the signature mutations at 126 and 294, which is the clan of Tara and the one to which I belong, everyone within the clan shares these two mutations, for the simple reason that Tara herself had these mutations and everyone in the clan is one of her direct matrilineal descendants. These two mutations have come down through the generations unchanged from the clan mother herself. But how many generations? How long ago did Tara live? That is where the additional mutations come in. Although roughly a third of people in Tara's clan have only these two mutations, the rest have additional changes. I have one extra mutation, at position 292, which makes my mDNA sequence 126, 292, 294. Other members of the clan have experienced more mutations. All these additional mutations *must* have occurred since Tara's time. Fortunately we know the mutation rate for the mDNA control region. It is approximately one change every 20,000 years. Since mutations happen completely randomly, not every line of descent from Tara will experience the same number of mutations. Some may be spared altogether and retain just the signature mutations at 126 and 294. Some maternal lines, like mine, will have been hit once since Tara's time, others more than once, some not at all. By working out the *average* number of additional mutations within the clan, we can then estimate how old the clan is, or, to put it another way, how long ago Tara herself lived. For her clan, the average number of additional mutations within the clan is almost exactly 0.85. With a mutation rate of 1 change per 20,000 years, the conclusion is that Tara lived 17,000 years ago.

Repeating the same calculations for the other six clans, we arrive at estimates for the ages of the other clan mothers. The clan with the greatest number of additional mutations on top of the clan mother's signature sequence is Ursula's. Hers is therefore the oldest of the seven clans. The average number of extra mutations in the clan is 2.75, and factoring in the mutation rate, this means that Ursula herself lived 45,000 years ago. Xenia is the next oldest at 25,000 years, Helena next at 20,000 years, then Velda and Tara both at 17,000 years, Katrine slightly younger at 15,000 years and finally Jasmine at 10,000 years ago.

Working out how long ago these women lived was a big step to discovering what their lives were like. Now I knew when they lived, could I discover where? I used three tests to find out. First, knowing the current whereabouts of the clan throughout Europe, I discovered where the clan was concentrated, reasoning that even after so many thousands of years, this might still be close to its origin. However, more important was to plot where the clan had accumulated the most additional mutations. The reasoning here was that the clan would have had longest to 'age' close to its origin, where the clan mother herself lived. To give you an example, the clan of Velda reaches its highest frequency in two places – northern Spain and among the Saami of northern Scandinavia. But it is far more varied, in the sense that it has accumulated far more extra mutations, in Spain than in Lapland. So I placed Velda herself in northern Spain, rather than in the far north of Norway and Sweden. Which brings me on to the third test. The location of the clan mother has to have been habitable at the time. In Velda's case, we know from the archaeological records that people were living in northern Spain 17,000 years ago, the date estimated from the additional mutations in the clan, but they were certainly not living in northern Scandinavia, which was under several kilometres of ice. By the same process, the other clan mothers were located to Greece (Ursula), the Caucasus mountains (Xenia), southern France (Helena), northern Italy (Katrine and Tara) and finally Syria in the Middle East (Jasmine).

With information from climate records and the archaeological evidence, I was able to find out what conditions must have been like for these women living at these locations at those times in the past. I discovered what their landscape was like, what sort of diet they had, what age they reached and, armed with this information, I wrote imagined lives for them.

Since they were published, the response has been both surprising and intriguing. My laboratory was overwhelmed by requests from all over the world from people who wanted to know from which of these women they were themselves descended. We had already repeated the process worldwide and found a total of thirty-six equivalent clans, so we could deal with requests from anywhere. We could not possibly handle this demand in the lab, if only because we were prevented from carrying on any commercial activities by the rules of our principal sponsors, the Wellcome Trust. So the University rapidly formed a spin-off company, Oxford Ancestors, to perform this service. But that is of only passing interest compared to the quite extraordinary underlying emotion that the concept clearly aroused. It proved to me that to many people, of which I am one, the idea that within each of our body cells we carry a tangible fragment from an ancestor from thousands of years ago is both astonishing and profound. That these pieces of DNA have travelled over thousands of miles and thousands of years to get to us, virtually unchanged, from our remote ancestors still fills me with awe, and I am not alone. One unexpected effect is that when two people discover that they are both in the same clan, they really do feel like close relatives, like cousins or siblings. I have seen this happen time and again, and indeed on the Oxford Ancestors website one of the most popular activities is discovering genetic relatives and then swapping personal information and often finding uncanny similarities of personality and circumstance. Even if this is all retrospective wisdom, after the test rather than before, the strength of feeling is very strong. There are even Jasmine parties organized by members of the clan.

I recently tested the DNA of our Vice-Chancellor, the executive head of Oxford University – I rarely travel anywhere without a DNA sampling brush – and discovered that he and I are not only in the same clan of Tara, but have exactly the same mDNA sequence 126, 292, 294. This means that as well as a common ancestor 17,000 years ago in Tara herself, we must share a much more recent maternal ancestor. I don't know who that is, but the point of the story is that, for better or worse, I feel now very differently about the Vice-Chancellor. So much so that, were we to have a severe disagreement, it would be hard for me to take it quite so seriously. It would be like arguing with my cousin.

A few years later, the same treatment became possible for the Y-chromosome. The details of the genetic changes were slightly different, and we will see how in a later chapter, but the principle remains the same. Whereas there are seven maternal clans which predominate in western Europe, there are only five principal paternal clans defined by the Y-chromosome. Each of these began with just one man, but for reasons that will become clear, it is much harder to know when and where they might have lived.

7

THE NATURE OF THE EVIDENCE

The collection phase of the Isles research project began ten years ago, in 1996, under the title of the Oxford Genetic Atlas Project. I obtained ethical permission to collect DNA samples from volunteers with the specific objective of discovering more about our genetic history. Over the next few years, my research team and I worked our way all over the Isles. We collected over 10,000 DNA samples and travelled over 80,000 miles by train, plane, boat, car and bus. Eventually I had to draw a line under the collection phase and concentrate on distilling some meaning from the thousands of DNA samples that now lay crowded in the lab freezers. We had been putting them through the analytical procedures more or less as they were being collected, converting the drab white threads of DNA into the sequences which would, or so we dearly hoped, hold the secrets of the ancient people of the Isles. Displayed on a computer screen they looked detached, dead – nothing like the talismans of ancient histories that I hoped they would become.

It took a lot of mental effort constantly to remind myself that every single one of these strings of letters and numbers represented the journey of an ancestor. A journey that at one stage almost certainly involved a sea crossing in a fragile craft to landfall on the Isles and an uncertain future. Fantastic though it sounds, it had to be true that each one of the thousands upon thousands of read-outs that flashed from the analyser to the computer in a fraction of a second had been carried across the sea in the cells of an ancestor. How could I get these mute listings to tell me their stories? How could I get them to sing? If only, I thought one day, I could read in the letters of the genetic code the language of the bearer. How wonderful that would be – and how much easier than the task that lay ahead. If, just by looking, I could recognize a Gaelic word or a Saxon spelling somewhere in the sequence of DNA letters. But the genes were stubbornly silent, oblivious to the tongues of their bearers.

Mathematicians have devised a whole array of statistical tests to sieve through DNA results, mechanically and without feeling. Indeed, most scientific papers on this kind of genetics spend at least half the time agonizing over what is the correct statistical treatment. It is necessary, if only to get results published, to know how to do this and fortunately we had in the lab several people skilled in the art. They, in particular Eileen, Jayne and Sara, put the accumulating genetic data through their paces. They ran Hudson tests, Mantel tests, distance-based clustering analyses, drew genetic matrices based on Fst and Nei's D, performed spatial auto-correlation tests and many more. Here are some of the results that came screaming out of the computer. It is a set of genetic comparisons from mitochondrial DNA between the four regions of the Isles.

Ireland/Wales	0.0726741243702487
Ireland/Scotland	0.0625191372016303
Ireland/England	0.1170327104307371
Wales/Scotland	0.0662071306520113
Wales/England	0.0980420127467032
Scotland/England	0.1023741618921030

You do not need to know what these mean, and I hope you do not want to. Even as I write them down, I can feel I am being drawn away from the real lives of these genes into some grey underworld where everything becomes a number. The genes are submitting to this cruel procedure, but they will never sing again. Now they are processed into numbers, with so many decimal places that they assume an importance way above their true worth. It feels as though I have handed them on to a windowless world which has severed any contact with the sea and the wind. Once a number is produced, something, perhaps everything, of value has been lost. Like so many tabulations, the numbers disguise individual stories of heroism and betrayal, triumph and defeat, and force them into bleak summaries. This is no way to treat our ancestors and you will be glad that I shall not insult them, or you, in this way again.

Since every ancestor was an individual, I was determined to treat the DNA sequences as individuals. Each one had, at some time, set off from some distant land and stepped ashore on the Isles, soaked with salt spray and red-faced from the cold. I decided that, if I possibly could, I would not treat these as anything but individual journeys undertaken with deliberate purpose and not to be grouped together in clumsy approximations. I covered the walls around my desk with photographs of the coast and the sea, of the Isles from the savage Atlantic to the smooth sands of Kent. Whenever I was tempted to revert to orthodox analysis I would glance upwards and remember to tread more carefully.

Finally, I had nearly 6,000 different pieces of genetic information from volunteers all over the Isles, each one linked to a geographical origin. By the time I came to write *Blood of the Isles*, I could add in another 25,000 genetic messages from among the customers Oxford Ancestors. I contacted colleagues whom I knew had similar genetic information from the Isles and from other parts of Europe. I trawled all the relevant publications for material. When I finally settled down to listen to the music of the genes I had over 50,000 DNA sequences to work with.

For two solid weeks over Christmas I sat down to get to know these details. Fortunately, the weather was awful. It was raining constantly and was very, very windy. I live close to a sea loch in Skye and, when the wind is strong and in the south-west, blowing straight in from the North Atlantic, it descends in howling gusts from the Cuillin Hills. These winds tumble off the main ridge of the mountains and roll down the loch, pulling the top layer of water into the air in spiral twists of spray. The oddest thing about these winds is their intermittence. The air is calm, windless and then you hear an approaching roar and it is upon you and so strong it is almost impossible to stand upright. Then, after five minutes' battering, it is gone. After another few minutes the sequence begins all over again. The alternating spells of chaos and calm can go on for hours. Hours well suited to going through thousands of sequences one by one, giving each one a different name and a different number. The coal fire burns well and the smoke is only very rarely forced back down the chimney.

This is how I got to know the data. Thanks to a mapping program written by a colleague, I could quickly place any selection of DNA sequences on a map of the Isles. As I did this I soon noticed that some DNA sequences were found in all parts of the Isles, while others were very localized. For instance, I had found one particular mDNA sequence in the clan of Tara four times in Skye, once in Lewis in the Western Isles and once in Glasgow – and nowhere else in the world. When I looked up in my records where on Skye the four people lived, I saw they came from different parts of the island. But all traced their maternal ancestry back to the Isle of Rona.

Drive to the north end of Skye past the eroded cliffs and pinnacles of Trotternish, high above the sea, and Rona is the low rocky island on your right, lying 5 miles offshore. It looks as if it is connected to the longer, higher island of Raasay to the south, but it is not. A hidden sea channel separates the two islands. Rona is deserted now, but once held a few crofting families who fished in the dark blue seas. Their houses have been abandoned and only the lighthouse, white against the rocks, is visible from Skye.

What must have happened on Rona to account for the unusual DNA I had found on Skye was a mutation, a slight change in the DNA of one of the ancestors. Silent, unnoticed and with no effect at all on the woman in whom this event had occurred, just one DNA base, one bead on the chain, had changed. The new sequence was unique, never seen before in the history of the world. If this woman had been childless or had only sons, it would have died with her. No one would ever know it had been created. But she must have had children, and at least one of her children must have been a girl for her mDNA to be passed on. Through this girl, or her descendants, this new sequence left the island of Rona and found a home on nearby Skye, where it still remains. From there, perhaps one of the daughters in the next generation went to live on Lewis while another travelled down to Glasgow.

I cannot tell exactly when this happened, but the journeys have been recorded by the genes of the descendants. I have not found this particular sequence of DNA letters anywhere else. Nor has anyone as far as I am aware. That doesn't mean it isn't there in other parts of Scotland, or Ireland, or Wales, or England. Just that we haven't found it. That is always the way, and always will be. We will never know everything there is to know about this new gene and what happened to it. We can only piece together something of its journey from the scraps of information that have both survived to the present day and that we have found in the cells of people we have tested.

This is a little story of one particular gene, a new version that has changed very slightly. If we ever do come across it again in the future, we will know it has travelled from Rona. It is a fragment, like a piece of pottery or a flint tool, and just as reliant on the twin necessities of survival and discovery as any archaeological remains. This is how I would build the genetic history of the Isles, by sifting through the thousands of fragments, trying to make sense of them. I would treat them as if they were the scattered shards of broken pottery and do what I could to understand what they meant. This was the point at which I decided to become a genetic archaeologist. I would work with fragments of DNA, perfectly preserved in the bodies of descendants, to reconstruct the travels of their ancestors with the same discipline that an archaeologist would use when excavating a site. Collect, examine, record, compare, interpret. In my mind's eye, even though they were in reality stored on my computer, I began to think of them as, literally, a pile of fragments, pottery perhaps or maybe coins. Yes, coins would be an even better metaphor. Through Chris Howgego, a friend and colleague from Oxford, I had been allowed to examine the Ashmolean Museum's collection of ancient gold coins from Britain. Many are over 2,000 years old but still fresh and lustrous, stamped with the image of an ancient tribal king or the stylized outline of a horse, a chariot wheel or an ear of wheat. I would set out to use the genes to interpret the past, just as Chris Howgego used his coins.

Of course they would tell very different stories. Coins and genes are not the same, but neither are they so very different. Both have to obey the rules of survival and discovery and, in the case of gold coins at least, both had been preserved virtually intact. Both bore inscriptions, either the name of a king or the sequence of a piece of DNA. A coin from a distant land discovered in Britain is a witness to a journey made, just as much as a fragment of mDNA must have made its way to the Isles at some time in the past in the cells of an ancestor. Coins have one thing that is conspicuously lacking in DNA and that is a date. Though the Iron Age coins in the Ashmolean Museum do not have a calendar date impressed on them, their date of manufacture can be worked out from the tribal chief whose image or inscription is on the coin. For example, several are inscribed with the letters CVNO, denoting that they were minted during the reign of Cunobelinus, King of the Catuvellauni some time between 60 and 41 BC.

With DNA we are not quite so fortunate. It does not bear a date stamp, but there is information on time depth to be had along the same lines that we have already used when working out how long ago the clan mothers lived. We can get an idea, though only an approximate one, of how long a group of mDNA or Y-chromosome 'gene-coins' has been in a location by seeing how they differ from one another. But it can be tricky, as we will find out.

I decided to excavate my first pile of 'gene-coins' from the results of the Genetic Atlas Project and I would begin with the mDNA, those fragments of history that have been passed down in the bodies of women. To begin with I did what anybody would do with a pile of gene-coins. I counted them. I had a total of 3,686. I imagined a large outline map of the Isles spread out in front of me ready for me to place each gene-coin in its correct location. But before distributing them, I sorted the large pile out into smaller groups of different types, depending on their clan. Once I decided on that course, the way forward became a little bit easier to imagine. I could now abandon the manacles of conventional statistical analysis and approach my reconstruction of past events as a genetic archaeologist. That meant, for one thing, that I could give up trying to know everything, and also stop pretending that giving answers to sixteen decimal places means anything at all.

Historians and archaeologists realized this a long time ago and make do with what they have, while being on the lookout for new material. Geneticists do not naturally think like that. They are not natural storytellers. If a geneticist does not get a watertight answer after an experiment or a survey of some kind, he or she will go back to the drawing board rather than risk saying anything that might be shown later to be wrong. If it is a survey and 1,000 samples have failed to produce a statistically significant result, our training tells us to say nothing and increase the number to 10,000, and if that doesn't work to 100,000. Having reinvented myself as a genetic archaeologist, I was free to do my best to tell the story of the Isles in my own way with what data are available. The account will never be complete, even if I were to double or quadruple the number of DNA samples. Also, even though I knew perfectly well that I would have to do the actual operations on my computer, the vision of the gene-coins as tangible objects which could be picked up and examined and then moved into position on a map was unexpectedly reassuring. I had rescued the project from the number-crunchers.

What did the pile of gene-coins look like? It was easy to decide on how to create the different piles. I would arrange them according to their maternal clan. I can tell the maternal clan of one mDNA sequence from the combination of mutations that it has. If I see the combination 126, 294 I know I am dealing with a member of Tara's clan. If the sequence contains 256, 270 this is the mDNA of an Ursulan, and so on. These mutations became the inscriptions on the gene-coins and the portraits changed from tribal chieftains to the rough profiles of the seven matriarchs, Ursula, Xenia, Helena, Velda, Tara, Katrine or Jasmine.

In my mind the action moved to a baize-covered table. I soon sorted the large pile into smaller ones, one for each maternal clan. In the largest of these clan piles I had 1,799 gene-coins with the profile of Helena. The next biggest was the 434 in Jasmine's pile, followed in sequence by 384 Tarans, 284 Katrines, 264 Xenias, 207 Ursulans and lastly 116 Veldans.

But there were still a lot of gene-coins that remained in the unattributed pile. I looked at the portraits and the inscriptions. These were of other matriarchs, not the Seven Daughters of Eve, but ones I still recognized. The most common were the gene-coins belonging to the matriarch Ulrike. There were 101 in all, only a few short of the Veldans. I had not included Ulrike as one of the what would then have been Eight Daughters of Eve because, in the research in Europe, the clan of Ulrike was considerably less frequent than the other seven in the regions we had surveyed, which were mainly the southern and western parts. As more information came in from Scandinavia and eastern Europe, we saw more and more members of Ulrike's clan. I've wondered since whether Ulrike should be promoted, as it were, into the select group of clan mothers.

But even with the Ulrikans now separated from the rest, there were still quite a few gene-coins in the pile. They were an exotic collection, from matriarchs all over the world. I stacked them together for now. There were ninety-seven in all. These, then, were the fragments with which to build the genetic history as told by women.

On the male side I had 2,414 Y-chromosome gene-coins from the Genetic Atlas Project and began to sort these into different piles according to their clans. Though the genetic details were displayed in a different form, the principle was the same. Each clan, of which there were five major ones in the Isles, traced a direct patrilineal line of descent right back to a common ancestor, the man who had founded the clan. In the Isles, these were the clans of Oisin (pronounced Osheen), Wodan, Sigurd, Eshu and Re.

Even though I knew full well that the gene-coins did not exist in reality, the concept gave me a lot of confidence. I began to relish the prospect of trying my best to interpret them and what they told of the past, rather than despairing as I had been up to then. Now, at last, I was mentally ready to launch into the final stages of the project. It was now as an archaeologist that I settled down to explore the Blood of the Isles.

8

IRELAND

The Irish landscape has often been compared to a bowl. A broad central limestone plain dotted by lakes and peat bogs and drained by sluggish rivers is surrounded by coastal ranges of hills and mountains. This upland barrier is only breached to any significant extent around the capital, Dublin. The total land area is 32,000 square miles (26,600 in the Republic and 5,400 in Ulster). The highest peaks, Lugnaquillia (926 metres) in the Wicklow Mountains south of the capital and Carrantuohill (1,041 metres) in Kerry, are on a par with the tallest mountains in Wales and England but well below many of the highest peaks in Scotland. In the west, the mountains thrust out long fingers into the Atlantic Ocean, creating a series of deep bays between, many of them now flooded river valleys. In the far south-west these rocky fingers are formed by parallel folds of sedimentary old red sandstone, like parts of northern Scotland, but further north in Galway, Mayo and Donegal, as well as in the Wicklow Mountains to the east, the rock is granite, the weathered remnants of once-molten magma forced to the surface by ancient movements of the earth's crust. On the eastern coast, facing Britain, the coastline is more orderly, without the drama or the dangers of the stormbound west.

As in the rest of the Isles, the landscape has been sculpted by ice. During the last glaciations, the ice covered only the northern half, extending as far as a line between Limerick in the west and Dublin in the east, but earlier Ice Ages enveloped the entire land in their frozen grip. The scouring of the central lowland plateau created the bedrock upon which the great peat bogs later grew and which, later still, provided the main supply of fuel for generations of rural households. The ice also ground the limestone base into a powder which formed the most important element of Irish soil. Without the glacial limestone powder to enrich it, the soil, made up of the weathering from older rocks like quartzite, granite and shale, would be infertile and unproductive like so much of the Scottish Highlands. But limestone gives it life, and thanks to this essential enrichment, and to the high rainfall, Ireland has thrived on its green pastures. Without the limestone, Ireland would not be the Emerald Isle, but the Brown.

On the 'Irish History' shelves of any high-street bookshop, the titles on display are dominated by the political struggles of the last hundred years. Books abound on the Easter Rising of 1916, alongside biographies of Michael Collins, Eamon de Valera and other heroes in the struggle for independence from Britain. A struggle which continues to this day, as Republicans strive to unite Ireland into the single nation it once was. As I write, in 2006, the intensity of the cycle of violence and recrimination has diminished considerably. The Irish Republican Army has renounced violence as the means of achieving its political aims and has destroyed its weapons. The shaky Good Friday agreement of 1998 staggers onward and, with luck, it will not be long before the devolved power-sharing government between the republican Sinn Fein and the opposing parties is re-established. The roots of this struggle go back a very long way and, though *Blood of the Isles* is certainly not a political history, it is as well to be aware of events which may have had some influence on the genetic patterns we are setting out to interpret.

The current struggles for Irish political unity and independence are but the latest stages in a chain of events that began over 800 years ago when, in the autumn of 1171, Henry II, the Anglo-Norman king of England, landed in Ireland to make sure that it did not become a rival Norman state to his own. Five years earlier, the ambitious Richard de Clare, Earl of Pembroke, himself an Anglo-Norman, had responded to an invitation by one of the numerous Irish kings, Dermot MacMurrough, whose lands in the Leinster had been seized by the High King, Rory O'Connor. This was a classic situation, seen many times before and since, where an invitation by a dispossessed or threatened king is used as a cover for invasion. De Clare, whose sobriquet 'Strongbow' adequately describes his attitude to conquest, seized the chance and established a secure foothold in Wexford and the south-east part of Ireland which faced his base in Pembroke, only 40 miles by sea. His military campaigns were extremely effective, thanks largely to the superior weapons he brought across. Heavily armoured knights, especially when mounted on horseback, easily overcome the local opposition armed only with light bows and spears.

Anxious that de Clare's success did not lead to the establishment of a rival kingdom, Henry arrived to impose his authority. This he did by granting Leinster to de Clare and County Meath to one of his own commanders, Hugh de Lacy, while at the same time forcing the remaining Irish kings into various forms of submission, including the obligation of giving forty days' military service and requiring Henry's permission to marry. From then until 6 December 1921, when three of the four provinces broke away from British rule to become the Irish Free State, Ireland's history and its fortunes were tied to England's. The name changed to Eire in 1937 and, finally, became the fully independent Republic of Ireland in 1948.

The occupation of Ireland by the English between these dates was never entirely convincing and oscillated between periods of calm indifference and others of turmoil and ruthless exploitation. The exclusion of Ulster from the Irish Free State was the visible residue of the Protestant Ascendancy which followed the defeat by William of Orange of James II at the battle of the Boyne in 1690. This well-remembered and, in Ulster, celebrated victory was the culmination of centuries of rebellions and uprisings within Ireland against the influence of England. In Ireland, not only the Gaelic lords but also the descendants of the Anglo-Normans frequently found their lands confiscated as, from the time of Elizabeth I, they were granted as plantations to favourites and adventurers. The policy was continued by Elizabeth's successor James I of England (James VI of Scotland), who encouraged the large-scale settlement of Ulster by lowland Scots. From the genetic point of view, what distinguishes this episode from what had gone before is that, instead of estates merely changing hands from one member of the aristocracy to another with little effect on the majority of the population, the plantation of Ulster imported tenant farmers and labourers from Scotland to work the land. The earlier Anglo-Normans had not as a rule imported their labour force, so we would not expect any genetic influence to be felt especially strongly. However, in Ulster we need to be aware of the possible effects of the plantations on the genetic patterns.

Before we leave the turbulent centuries of Irish history, there is one more episode that we must not forget. So far we have only mentioned immigration into Ireland, by Anglo-Normans at first and then through the plantations. But these are numerically dwarfed by the departures. Religious intolerance and persecution from the sixteenth century onwards, closely coupled to land seizure, drove many Catholic landowners abroad, mainly to France and Spain. Though doubtless traumatic for them, these exiles did not really affect ordinary Irish agricultural workers, for whom life continued much as before, though the land was under new ownership. However, in the nineteenth century, Irish emigration on a large scale began in earnest.

In the first decades of the century, agricultural prices fell, estate rentals declined, investment in the land was reduced to a trickle, and the rural population grew. Whatever the ultimate causes of this cycle of economic decline, the effects on the rural poor were catastrophic. Reduced to almost complete dependence on the potato as the staple crop, the countryside was decimated when the crop was infested with the potato blight and rotted in the ground. During the Great Famine of the mid 1840s, thousands died of starvation or of the infectious diseases which swept through the malnourished population. Though many thousands died, thousands also made their escape. Ireland's mid-nineteenth-century population of 8 million began a steady decline that has only very recently stabilized at 4.1 million in the Republic and 1.7 million in Ulster. The desparate diaspora of the Irish saw massive immigration both to Britain and to the New World, especially the United States. Today, there are far more 'Irish' genes abroad than there are in Ireland itself.

Though Ireland is not yet united into a single political state, the poverty and suffering which suffuse all accounts of the history of the last centuries cannot be equated with Ireland today. The economy is transformed. The bars and cafés of Dublin are as lively and as sophisticated as anywhere in Europe. There is a tangible feeling of optimism in the air wherever you go. Though we will have to wait to see how much of the turmoil of past centuries is remembered by the genes, I suspect the main effect will be of emigration and the dispersal of Irish genes around the globe. Now that the future of Ireland as an independent country is looking so good, this is the time to move the sad centuries to one side and examine Ireland before the day when Henry II arrived to begin the English occupation. That is where we must seek to interpret the patterns of the genes. What do we know of these earlier times?

The appeal that Dermot MacMurrough made to Richard de Clare to come to his aid, the appeal de Clare used as an excuse to invade, is a clear indication of the state of affairs in medieval Ireland – the struggle for dominance of one minor king against another. It is so very typical of the middle stage of evolution of any modern society and one that is only too visible in other parts of the world. Except that in those places, like Afghanistan or unstable African countries, these men are not dignified with the title 'king' but denigrated as 'warlords'. In Ireland during the first millennium AD there was a constant struggle for dominance between different minor kings. According to one source, there may have been 150 of them at any one time, lending some credibility to the common Irish boast that they are all descended from lines of Irish kings. This may be something we can test as it could be visible in the Y-chromosome gene pool by what has come to be known as the 'Genghis Khan effect'.

A few years ago, researchers from Oxford found a Y-chromosome that was very widespread throughout Asia, more or less within the geographical limits of the Mongol Empire. Finding a particular Y-chromosome with a specific fingerprint across such a wide area is highly unusual. Y-chromosomes are generally much more localized. The explanation, which I think is the correct one, is that this is the Y-chromosome of the first Mongol emperor, Genghis Khan, who lived in the first half of the thirteenth century. Not only is the Y-chromosome fingerprint geographically dispersed, it is also very common. In Mongolia, for example, 8 per cent of men have inherited the Genghis chromosome. If you compute the number of men who carry this Y-chromosome throughout Asia, and occasionally on other continents, then it comes to a staggering 16 million. Even a cursory glance at Genghis Khan's methods in warfare is enough to understand the genetic mechanism. On conquering an enemy's territory he would kill all the men, then systematically inseminate all the good-looking women – he left his commanders strict instructions on that point. When he died, the custom of patrilineal inheritance ensured that his empire was distributed among his sons, and their sons. Thus his Y-chromosome increased with each generation of male descendants, who inherited not only a portion of his wealth but also, presumably, his attitude to women. Though we have no historical records of men with quite such sexual predominance in the Isles, the confusion of minor kings is just the sort of condition where one might expect to discover the Genghis effect.

It was not all chaos in Ireland. Some kings managed to exert sufficient authority to stake a claim to the title of High King and to be installed at the sacred site of Tara, about 20 miles north of

Dublin. Though none of the High Kings ever managed complete dominance over the whole island, some had a very good try and this may well be reflected in an Irish Genghis Khan effect. While such behaviour may rearrange the genes of Ireland, or anywhere else in the Isles for that matter, it is however only a rearrangement. While the Genghis effect will mean that one, or a few, Y-chromosomes may prosper at the expense of others, no amount of Khan-like behaviour can actually create new Y-chromosomes. And it has no effect whatsoever on the maternal lineages, traced by mitochondrial DNA. These will persist whatever the kings get up to.

Peering further back into the Irish past, what can we see that needs to be taken into account? Though it was Ireland's misfortune to be occupied by the English for so long, it entirely avoided being conquered by the Romans, which large swathes of Britain did not. Ireland was very lucky to escape. The Roman historian Tacitus, in his account of the campaigns of his father-in-law Agricola, tells how the great general seriously contemplated an invasion. In his fifth year of campaigning in Britain, in AD 88, Agricola brought his army and his ships to Galloway in southeast Scotland, only 20 miles across the sea from Ireland. Such were the inaccuracies in the geography of the day that Agricola believed that Ireland was midway between Britain and the Roman province of Spain. So he could see the tactical advantages of including Ireland within the Empire. He had received favourable reports about the character and way of life of the inhabitants and of the soil and climate. To a Roman they did not differ much from the British, whom he had successfully subdued during the previous five years. Tacitus wrote that he often heard Agricola say that Ireland could be conquered, and held, with a single legion supported by a modest force of auxiliaries. Agricola even took the precaution of befriending a minor Irish king who had been exiled in case the opportunity to use him should arise. In the end he decided against an invasion. Tacitus does not say why and we can only guess. But that he had serious intent is certain.

One negative consequence of this lucky escape was that there are no written histories of Ireland from the Roman period. Not until the arrival of early Christians in the fifth century AD, and of St Patrick in particular, did written accounts, however unreliable, begin to appear. St Patrick himself is credited with the authorship of the earliest documents in Irish history, written in Latin: the *Confessions*, which defines and defends his mission, and one other, a short letter excommunicating the soldiers of a British chieftain who had murdered some of Patrick's converts. Neither account throws much light on life in Ireland at the time – nor was that the intention. None the less, the beatification of St Patrick and his emergence as the supreme cult figure, which in many ways he remains to this day, did lead to further accounts of his life and his Christian mission by later authors. The early ninth-century *Book of Armagh* is the culmination of these and it established the primacy of the See of Armagh in the Irish Church.

The three centuries following St Patrick's death in AD 493 are rightly regarded as a golden age in which Ireland became one of the most important religious centres in the whole of Europe. It was from Ireland that missionaries set out to convert the pagan tribes of northern Britain, establishing Columba's monastery on Iona in AD 563 as a stepping stone. From Iona the monastery of Lindisfarne, off the Northumbrian coast, continued the mission to the eastern side of Britain. Irish missions to continental Europe were equally successful and St Columban, not to be confused with Columba, founded monasteries at Luxeuil in the Vosges mountains of eastern France and at Bobbio in the hills of the northern Apennines in Italy in the late 600s. Other Irish monks sought the opposite – a life of austere contemplation – and their search for solitude took them to increasingly remote destinations. They found what they were looking for on rocky islands off the west coast of Ireland, in the Western Isles of Scotland and even as far north as Iceland. Their journeys across the wild seas are all the more remarkable for having been undertaken not in well-constructed galleys but in curraghs – light boats with only shallow drafts and made from wooden spars covered in tarred animal hide.

But still, despite the intensity of religious devotion and scholarship, the written accounts are more or less completely bare of historical content. Even so, it is fairly clear that the reputation of these early saints was linked to the fortunes of the political dynasties to which they became attached. An association with the cult figure of St Patrick himself was the ultimate claim to authority and influence, and the opportunity was not overlooked by the first of the invading Anglo-Normans. In 1185 one of these barons, John de Courcey, arranged for the 'discovery' of St Patrick's remains at Downpatrick and their removal to Armagh. This is very reminiscent of the fabricated discovery and ceremonial reburial of King Arthur's bones by Edward I at Glastonbury a century later. Clearly, in the medieval period, any association with long-dead cult figures could be used as a claim for historical legitimacy.

Rather as in England, with the loosely based fiction of Geoffrey of Monmouth, history was written for a purpose. The Irish equivalent of Geoffrey's *History* was the *Leabhar Gabhála*, the *Book of Invasions*, compiled from earlier writing in the late eleventh century. Even though, just like Geoffrey's *History*, it is a clear attempt to link Irish history to the familiar events of the classical world, it managed to create a compelling narrative for the origins of Ireland and of the Gaels which became extremely influential as an origin myth for the Irish. And it still is. However accurate or inaccurate it may be as a record of Irish origins, we must still bear it in mind when we sift the record of the genes. Deeply held origin myths, however richly embroidered, have a habit of being right.

Although the *Leabhar Gabhála* was doubtless compiled by Christian monastic scribes, in common with the written versions of other rich mythologies in Ireland, there was no conflict or contradiction in recording the pagan myths of their native or adopted land. The phenomenal success of Irish Christianity owed a great deal to the sympathy it showed to ancient traditions and rituals and to their preservation in written form. The Irish Christian monks became the conduit of ancient knowledge, the *filidh*, and their success lay in their ability to create a seamless continuity between the rich mythical traditions of pagan Ireland and full-blown Christianity.

From our point of view, the *Leabhar Gabhála* chronicles four mythical phases of immigration. As you can imagine, all four involve great battles and heroic struggles as each wave of new arrivals ousts the former occupants. The last of these phases was the invasion of Ireland by the Gaels, bringers of the language and the alleged ancestors of today's Celtic population. Indeed the principal purpose of the *Leabhar Gabhála* is to explain the presence of the Gaels in Ireland.

According to the *Leabhar*, the Gaels were descended from the sons of Mil, also variously known as Milesius and later by the, perhaps significant, epithet of Míle Easpain, or the 'Soldier of Spain'. Mil was killed on an expedition to avenge the death of a nephew who had been killed by the Tuatha Dé Danaan, the previous occupiers and masters of Ireland. It was left to Mil's three sons, Eber, Eremon and Amairgen, to defeat the Tuatha and conquer Ireland. When the brothers could not agree on the division of the island between them, Eber was killed by Eremon, who became the first High King to reign at Tara. Mil's wife, Scota, was also killed in the expedition and the Gaels of Ireland, considering her to be their ancestral mother, called themselves Scots for that reason. Certainly the Romans referred to them as *Scotti* as well as the more familiar *Hibernii*.

According to legend, the ultimate ancestor was one Fennius Farsa, a Scythian king who lost his throne and fled to Egypt. Ancient Scythia was located north of the Black Sea in what is now the eastern Ukraine, between the two great rivers, the Don and the Dnieper. Once in Egypt his son, Nial, married the pharaoh's daughter, and she had a son, Goidel. The whole family was banished from Egypt for refusing to join in the persecution of the children of Israel and wandered throughout northern Africa, finally crossing the Pillars of Hercules to settle in Spain, where they prospered.

Many years later, from a watchtower on a cliff top, one of Goidel's descendants, Ith, saw a land far off across the seas that he had not noticed before. 'It is on winter evenings, when the air is pure, that man's eyesight reaches farthest,' explains the account of the vision in the *Leabhar Gabhála*. Although it is quite impossible ever to see Ireland from Spain, Ith wasn't to know this and he set sail with ninety warriors to explore the newly sighted country. He arrived at the mouth of the River Kenmare, one of the deep indentations in the coast at the extreme south-west of Ireland. From there, Ith tracked northwards until, at last, he encountered the Tuatha Dé Danaan, the race who inhabited Ireland. The meeting went well at first until the Tuatha began to doubt Ith's motives for sailing to Ireland and, from his fulsome descriptions of the climate and the fertility of land and sea, suspected that he intended to invade. They killed Ith, but spared his companions, who then returned to Spain with their leader's body. Ith's uncle Mil vowed to avenge his nephew's murder and set sail with his eight sons and their wives, accompanied by thirty-six chieftains, each with a ship full of warriors. With his sons at his side he defeated the Tuatha. Mil was himself killed in the battle, but his sons survived. The defeated Tuatha Dé Danaan also chose their name from their own ancestral mother, Dana. The Tuatha were a race of gods, each with their own special attributes and each as colourful as any gods of the classical Greeks. After their defeat by the Milesians, the Tuatha Dé Danaan fled to the Underworld and established a kingdom beneath the ground – a kingdom from where they were still able to harass their conquerors by depriving them of corn and milk, eventually forcing an agreement which divided Ireland into upper and lower parts and in which the Tuatha Dé Danaan are to this day the guardians of the Underworld.

In their own conquest of Ireland, the Tuatha Dé Danaan had ousted two groups of earlier occupants – the Fir Bholg and the Fomorians. After their defeat the Fir Bholg, a race of pre-Celtic humans, were banished to the Aran Islands in Galway Bay. Unfortunately, the *Leabhar Gabhála* does not say where the Fir Bholg had come from. The implication is that they had been there all the time. In this respect, the Fir Bholg resemble myths in other parts of the Isles about a race of aboriginal inhabitants, usually described as being short and dark, who were subsumed by later 'waves' of Celtic arrivals.

The Fomorians, being divine like the Tuatha Dé Danaan, were altogether more difficult to defeat. Led by the terrifying Balor of the Baleful Eye, whose gaze alone caused instant death, the Fomorians were a race of demons. Balor's one weakness was the prophecy that one day he would be slain by his own grandson. Despite hiding himself away on Tory Island off the Donegal coast and keeping his daughter away from men, she nonetheless became pregnant and bore triplets. Balor threw all three of his grandchildren into the sea, but one, called Lugh, survived. He grew up to lead the Tuatha Dé Danaan against the Fomorians and, in fulfilment of the prophecy, killed his grandfather Balor with a slingshot through his one, baleful, eye.

Lugh went on to feature in the best-known myths of the Ulster Cycle, which records the continual struggles of the *Ulaid*, the Ulstermen, against the neighbouring province of Connacht. He becomes one of the many suitors of the notoriously promiscuous Queen Medb. No man could rule in Tara without first mating with Queen Medb. Fiercely competitive, as well as promiscuous, Medb's rivalry with one of her many consorts, the King of Connacht, leads into the most famous of all Irish myths, the *Taín Bó Cúalnge*, the *Cattle Raid of Cooley*. At first sight, cattle raiding might appear to be too prosaic a topic for a major myth, but remember that cattle were as much a badge of prestige as gold or jewels. Cattle raiding was an endemic occupation in Ireland as elsewhere in the Isles – and it was a failed cattle raid which led indirectly to the defeat of the giant Albion by Hercules.

The *Taín Bó Cúalnge* begins as Medb and the King of Connacht, in bed one night, decide to compare their material assets to resolve which of them is the richer. One matches the other until only a single item separates them. Ailill, King of Connacht, is the owner of a magnificent white-horned bull Findbennach, something that Medb does not possess. In vain she searches her own lands for a beast of comparable magnificence.

Then she hears of a great brown bull, Donn, and arranges to borrow it from its owner. Things start to go wrong when her soldiers brag that they could have seized the bull with or without the consent of the owner, who, overhearing their boasting, cancels the arrangement and hides the bull. Queen Medb decides on a disproportionate response and invades Ulster, precipitating a lengthy war between Connacht and Ulster. To escape the fighting, the great bull Donn is sent to Connacht for safety but, unwisely, bellows loudly as he arrives in his new home. His bellows disturb Findbennach, and he challenges Donn to a duel to the death. Their fight takes them all over Ireland until Donn eventually manages to impale his rival on his horns. Though he wins the contest, Donn does not survive to enjoy his victory and dies from exhaustion.

Forgive me for relating the *Taín Bó Cúalnge* at such length. It portrays the intense feuding and futile rivalry between the rulers of the different parts of Ireland more vividly than any purely historical account. And these are rivalries that might just have a genetic effect. The *Taín* also involves another super-hero of Irish myth, Cú Chulainn. The son of Lugh, slayer of Balor of the Baleful Eye, he is fostered as a child by two other heroes with somewhat exaggerated attributes. The first, Ferghus mac Roich, has the strength of 700 men and a prodigious appetite. He can consume seven pigs, seven deer, seven cows and seven barrels of liquor at one sitting – and he requires seven women at once to satisfy him. When Ferghus is killed, while bathing with Queen Medb and thus temporarily distracted, another hero, Conall Cernach, takes over as Cú Chulainn's foster-father.

Conall is the great champion of Ulster, who boasts that he never sleeps without the head of a Connachtman (severed presumably) resting beneath his knee. After foster-parenting like this, no wonder the boy grows up to be a super-hero. Naturally he is brave, beautiful, strong and invincible, and his chariot, helpfully, possesses an invisibility blanket to be used in the heat of battle. His weapons too are magical. His barbed spear, *Gae Bulga*, never wounds, only kills. In the war between Ulster and Connacht precipitated by Queen Medb's cattle raid, he kills vast numbers of her soldiers single-handed. His technique in battle is to transform himself into a berserk demon. His body spins round within his skin, his hair stands on end and one eye disappears into his head while the other bulges enormously. Small wonder his enemies are driven mad with terror.

Cú Chulainn is destined for a short though glorious life. By accidentally eating dog flesh one day, he breaks a vow that he made when a young man. His power drains away at the height of battle, his weapons fall at his feet and the Morrigan, a coven of divine destroyers, perch on his shoulder in raven form. Realizing he is no longer invincible, the Connachtmen pluck up the courage to approach and cut off his head.

As well as powerfully portraying the intense rivalries in early Ireland, the myths and heroes of the Ulster Cycle still exert their effect today. It is no coincidence that a bronze statue of Cú Chulainn, cast in 1916, the year of the Easter Rising, stands today in the hall of Dublin's main Post Office, which was itself the principal battleground of the Rising and the place where the Republicans held out for longest against the British. Myths are powerful things. And they often contain more than a grain of truth. But as well as these rich origin myths, there is an abundance of solid, archaeological evidence of Ireland's past.

The first signs of human occupation in Ireland are at Mount Sandel, situated on a bluff overlooking the River Bann in County Antrim. The site at Mount Sandel has all the signs of containing a substantial dwelling, with large numbers of round holes dug into the ground. Though these holes were filled by debris long ago, their outlines are clear. These are post-holes and they were dug to hold in place the wall timbers of a house. The wood itself has long since rotted, but the holes remain and, from their arrangement, the outline shape of the building can be made out. The house was round and, from the angle of the post-holes, the timbers were inclined inwards, suggesting a structure resembling a large tent 5.5 metres in diameter. Unsurprisingly, nothing remains of the roof, but plenty of later structures are known where the space between the roof timbers was covered by skins, twigs and reeds and there is no reason to think Mount Sandel was any different. Within the house there is a large square hole, probably a central hearth, and outside there are further pits, probably used for storage.

The large numbers and the variety of food remains found at Mount Sandel certainly suggest that it was used as a base camp throughout the year. There are hundreds of salmon bones, which show that the site was occupied in the summer when the salmon, fresh from the sea, pushed upstream to their spawning grounds. Huge numbers of hazelnuts and the seeds of water lilies, wild pear and crab apple show that the site was used during the autumn harvest of wild forest food. The remains of young pigs, which are born in the late autumn, are the sure sign of winter occupation. Overall, it looks as though this was an almost permanent base from where the occupants ranged over a 10-kilometre radius to cover the river, the estuary and the coast. Everything they needed was within a two-hour walk.

Carbon-dating of animal and fish bones found at the site reveals that Mount Sandel was occupied about 9,000 years ago, making the dwellings the oldest houses in the whole of the Isles. There are plenty of flint tools at the site and they are dominated by the small sharp flakes known as *microliths*. These were struck off a central core of flint and then fashioned for a number of different uses. Some were square in shape, with one or two edges finished sharply for use as cutters and scrapers. They were used for slicing animal skins and then stripping away the subcutaneous fat ready for drying and making up into clothing. Others were shaped to a sharp point for making holes in skins in preparation for sewing with sinews removed from the hind legs of deer. There are also hundreds of small flakes, some no more than a centimetre long, deliberately sharpened along one or two edges for use in composite tools such as arrows and spears. Like the roof timbers of the houses, the wooden shafts of these implements have rotted away so that only the stone remains.

The date of 9,000 years ago and the style of the material remains place Mount Sandel squarely in what is referred to as the Mesolithic period, otherwise known as the Middle Stone Age. The occupants may even have been exact contemporaries of the 'younger' of the Cheddar Men. Archaeologists divide the Stone Age into three phases. The oldest – the Palaeolithic or 'Old Stone Age' – covers the period from when the very first stone tools were discovered in Africa, at least 2 million years ago. They were not made by our own species, but by other types of archaic humans long since extinct. Our own species, *Homo sapiens*, does not make its appearance until about 150,000 years ago and the arrival of our ancestors in Europe about 45,000 years ago marks the beginning of the final phase of the Old Stone Age – the Upper Palaeolithic.

This phase lasted until the end of the last Ice Age, 13,000 years ago. The period between the end of the last Ice Age and the adoption of agriculture is known as the Mesolithic. Each phase is linked to a particular fashion of stone tool, and the microlith is the typical style of the Mesolithic. It is very much smaller and more refined than the larger flints of the preceding Upper Palaeolithic. Even so, the boundaries between the different phases are very fluid. For example, the scrapers of the Mesolithic are very similar to the scrapers of the Upper Palaeolithic.

At the time when people were living at Mount Sandel, the whole of the Isles was still connected to continental Europe. This does not mean the inhabitants of Mount Sandel did arrive overland, only that it was possible to do so at the time. The ice had begun to retreat 4,000 years before the main occupation of Mount Sandel, and the colonization of the Isles by the earlier Cheddar Man and his contemporaries had begun at least 3,000 years before. However, the earth wobbled once again in its orbit and there was a sudden and severe 'cold snap' between 11,000 and 10,000 years ago, which may have forced the human occupants back down south and cleared the Isles once more. The boundary of the ice, which had retreated to more or less its present latitudes, began to spread south again. The sea was frozen right down to northern Spain and the plains of northern Europe reduced once again to barren and inhospitable tundra. But, very fortunately, this cold phase – known as the Younger Dryas – lasted only for about 1,000 years. At the end of the cold snap the earth began to warm up very suddenly and humans could once again resume the occupation of northern Europe, this time for good – or at least until the present day. At first the landscape was bare of trees, rather like parts of northern Scandinavia today. Large herds of reindeer and wild horse roamed across the open plains once more. By the time Mount Sandel was occupied, the landscape was filling with trees as the temperatures rose. This warming was not a gradual process: the temperature literally shot up from bitter cold to very mild within less than a century. Around 9,500 years ago the average temperature was as high as, or even higher than, it is today.

The sea ice retreated way beyond the Shetland Isles and the sea level rose again as the ice melted. First Ireland was separated from the rest of the Isles at around 8,500 years ago. That put a stop to the colonization of Ireland by some land animals and explains why there are no moles, lizards or snakes in Ireland. That is, of course, unless you prefer to believe, in the case of snakes, that it was St Patrick himself that banished them. These animals, though, did have time to establish themselves in Britain before it was eventually cut off from the European mainland by the rising water levels in the North Sea 500 years later, about 8,000 years ago.

By now the Irish landscape had changed from tundra to an open forest of birch trees. As the temperature continued to rise, this open woodland slowly changed to a thicker cover of hazel and, by about the time the Isles became completely severed from the rest of Europe, they were covered in a mature forest of elm, lime and oak. The herds of large mammals moved north if they could, but in Ireland their way was barred by the sea. Many, including the magnificent Irish elk, with antlers some 3 metres across, became extinct. They were replaced in the now dense forests by wild pig, red and roe deer and the aurochs, the ancestor of modern domestic cattle, and by a host of smaller mammals like squirrel and pine marten. From the remains at Mount Sandel and other Mesolithic sites, it seems that anything that moved risked being roasted on the campfire. The ideal places to live were near rivers, such as at Mount Sandel on the River Bann, or by the sea. Here you could have the best of both worlds. Fish and shellfish from the sea and rivers, hazelnuts, pork and venison from the forest. Not a bad life at all. All the best shoreline sites accumulated huge mounds, or *middens*, of discarded shells built up often several metres high.

From the overall size of individual Mesolithic sites, archaeologists estimate that the number of people occupying them was quite low, possibly just single nuclear families. There was not the same need to join together in hunting bands of twenty or so as there had been in the colder, tundra phases. Then the main prey had been the herds of large and dangerous animals like bison, which called for organized ambushes and teamwork among the hunters. Neither was there any need to move over large distances to keep up with the herds as they migrated from summer to winter feeding grounds. Though many Mesolithic sites that have been found were obviously temporary, used for just a few days, others, like Mount Sandel, were occupied for long enough to make it worthwhile building the timber-framed houses.

Though the inhabitants of Mount Sandel were certainly hunter-gatherers, they were not above manipulating the environment to make life easier. They deliberately created open glades within the forest to encourage hazel trees to grow. They did not need to fell the mature elms and oaks to do this, but merely to ring-bark them and wait for them to die and be blown over. By stripping away a continuous band of bark from around the trunk, the capillaries that carry water to the leaves are disrupted and the tree begins to die. The next winter storm may blow it to the ground. The unremarkable life of the Mesolithic hunter-gatherers continued at Mount Sandel and elsewhere in Ireland for thousands of years, leaving little trace on the landscape and few permanent signs, shell middens apart, for archaeologists to follow.

Meanwhile on continental Europe radical changes were under way. From modest beginnings in the Middle East, farming was beginning its unstoppable march towards the Isles. Ten thousand years ago in the Fertile Crescent, in that part of what is now Syria and northern Iraq that is drained by the Tigris and Euphrates rivers, people had learned how to cultivate wild grasses and how to replace hunting with domestication. Farming ushered in the New Stone Age – or Neolithic, to distinguish it from the hunter-gatherer Mesolithic – with a whole range of new stone implements for farming. They also made pottery. The invention of agriculture seems such a small change in the tactics of subsistence, yet it has led to the complete reshaping of the world into its modern form. Whole books have been written about this, and I will resist the temptation to go off at a tangent, restricting myself instead to the implications for our remote ancestors, and for the gene patterns that await our scrutiny.

Carbon-dates from farming sites and the comparison of different pottery styles show that agriculture spread through continental Europe by two principal routes. The split probably came as the first farmers reached the Balkans and the lower Danube from Turkey around 8,500 years ago, about the time that Ireland finally separated from Britain and the residents of Mount Sandel were tucking into yet another bowl of limpet soup. One group of farmers headed north to reach the great Hungarian plains, then, after a thousand-year pause, moved rapidly north and west along the major river valleys of the Oder and the Elbe towards the Baltic and the North Sea. They needed to clear thick forest to make enough space for cultivation. This they did by ring-barking and burning the dead trees and undergrowth, thereby fertilizing the soil with ash. By 7,000 years ago they had reached northern France, southern Belgium and The Netherlands.

Meanwhile the other group moved along the Mediterranean coast of Italy, southern France and Iberia. By 7,500 years ago they had reached the Atlantic coast of France. At each point along the way, in the forest and on the seashore, each group of farming pioneers encountered the earlier Mesolithic inhabitants, but there is no archaeological evidence that their interactions were anything but peaceful. Just as in Ireland, the highest density of Mesolithic hunter-gatherers was around the coast, rather than in the dense inland forests. In several places, particularly around the coast near Lisbon in Portugal, Neolithic farming communities lived fairly close to Mesolithic settlements and carbon-dating shows that both were occupied at much the same time. However, the newcomers chose sites a little way away from the estuaries favoured by the hunter-gatherers, instead setting up camp inland on higher ground between the main river valleys. As they were not competing for the same living space, this reduced the potential for conflict.

In Ireland the same process of peaceful co-existence seems to have accompanied the arrival of farming communities. There were thriving Mesolithic settlements all around the coast, some of which, like Sutton in County Dublin, had been occupied for long enough to accumulate enormous middens of discarded shells over 100 metres long. They certainly would not have thrown in the towel as soon as the first farmer paddled round the coast. At Ferriter's Cove on the Dingle peninsula in County Kerry, the presence of polished stone axes – which, like pottery, are a reliable signal of the Neolithic – among the otherwise Mesolithic remains at this shoreline site, shows that the hunter-gatherers were in contact with farmers. Cattle bones at the site also show this inter-action. So, in Ireland, just as elsewhere in Atlantic Europe, the transition to farming from hunter-gathering was gradual and piecemeal and did not necessarily involve sharp changes in the make-up of the Irish population.

These signals of the arrival of the Neolithic in Ireland are small and subtle, noticed only by the professional archaeologist. How different, then, from the gigantic stone structures that also appeared in Ireland 1,000 years later. These are the jewels of Irish archaeology, drawing tens of thousands of visitors each year to stand in awe and reflect on the grandeur, the construction and the purpose of these magnificent structures. The Oxford archaeologist Barry Cunliffe has studied megalithic structures in the Isles and also in Brittany and along the Atlantic coast of France and Iberia. Rather than a phenomenon solely linked to the Neolithic and the spread of farming, Cunliffe traces their origin to the shell middens of Mesolithic Portugal. Within the piles of shells accumulated over centuries on the banks of the River Sado, excavations have found human remains that have all the appearance of being deliberate ritual burials. The middens are enormous, some over 100 metres in diameter and several metres high, and within some of them over 100 burials have been discovered. Further north, on the southern coast of Brittany, later dated midden graves have been found lined with stone. In others, bodies were buried with personal ornaments such as drilled sea shells and stone pendants. Traces of red ochre show that, like the Red Lady of Paviland, the bodies were covered in this pigment, the purpose of which may perhaps have been to restore the flush of health to a lifeless corpse.

Barry Cunliffe sees a natural progression from these shell-midden graves to the two earliest styles of Neolithic monumental architecture: the long barrow, where soil has taken the place of shells, and the passage graves. And it is the passage graves of the Boyne valley in Ireland that visitors flock to see. Although there are over 230 passage graves in Ireland, it is the tombs at Knowth, Dowth and Newgrange that, deservedly, command the most attention. All three are roughly the same size, 85 metres in diameter and 11 metres high. These dimensions may be similar to the shell middens which provided the archetypal design, but the effort put into their construction was phenomenal. The stone-lined passage and the tomb that lies at its end involved the quarrying, transportation and setting in place of over fifty giant stone slabs, some weighing more than 5 tons. Once the tomb was in place, the whole structure was covered in the gigantic mound, which is made up of more than 200,000 tons of rocks and earth.

At Newgrange, the narrow passage which leads to the tomb itself is 25 metres long. It was aligned in such a way that the light of the rising sun at the midwinter solstice shone directly along the passage on to an intricately carved triple spiral motif on the opposite wall of the central tomb. The Knowth mound, about a kilometre to the north-west, contains not one but two passage graves, as does the Dowth mound to the east of Newgrange. Around these massive tombs are other smaller tombs and numerous standing stones. Carbon-dates of organic remains found buried within the mound date the construction of Newgrange at about 5,000 years ago, well after the dates for Ireland's first unambiguously Neolithic site at Ballynagilly in Ulster.

It is only natural to imagine that these gigantic structures, and the complex and mysterious social rituals which their presence suggests, must have been brought about by a wave of new arrivals to Ireland. Yet the clear link to similar, even if not identical, structures along the entire Atlantic coastline, coupled with the early genesis of these structures in the middens of the Mesolithic, could equally well mean that these impressive megaliths are actually one step along the path of a continuous development of monumental architecture along the entire Atlantic fringe from Iberia to the Isles. That is definitely something to bear in mind when we contemplate the living archaeology of the genes.

Before we do that, let us pause to examine the archaeological evidence in Ireland of what nearly every popular account refers to as the arrival of the Celts. We have looked at the linguistic evidence elsewhere, but what is there to see among the material remains exposed by the trowels of the excavator? Even supposedly authoritative popular world histories describe 'the Celts' as loose mobile units of warriors, on the move and destroying all in their path. The main archaeological evidence comes from the beautifully fashioned and distinctive metalwork associated with, first, the Hallstadt and then the La Tène cultures of central Europe. Certainly these have been found in many parts of Europe, including Ireland. But to take this as proof of a large-scale movement of people into Ireland is surely, in the absence of other compelling evidence, even more of a risky assumption than that the spread of agriculture can only have been accomplished by wave after wave of Middle Eastern farmers.

The 'Celtic' artefacts that have passed the test of survival and discovery are, almost without exception, high-value goods that, like a Rolex watch or a Cartier necklace today, are at least as likely to be given as a gift as to be worn by the original owner. To find a La Tène brooch in an Irish bog is no proof that a central European put it there. And, like a fake Rolex, just because a piece of jewellery looks like the original, it doesn't stop it being a copy. In fact, artefacts in the La Tène style are rare in Ireland. Many of them have been examined by the Irish archaeologist Barry Raftery, who is convinced that, far from being made in central Europe, they were actually manufactured in Ireland itself. We continually underestimate the skill and capabilities of our ancestors. Why should it come as a surprise that an Irish goldsmith could learn a new, fashionable continental style? It seems to me that the constant tendency to interpret past events in terms of movements is completely the wrong assumption. Surely the correct starting point is to assume that our ancestors were sufficiently resourceful and skilful to pick up virtually any skill. But to find out we need to look at the DNA.

9

THE DNA OF IRELAND

A good reason for choosing Ireland as the starting point of our genetic tour of the Isles is that, unlike in Britain, a concerted research programme into Irish cultural and genetic history has already been running for some years, organized through the auspices of the Royal Irish Academy based in Dublin. In Britain an even more ambitious millennium initiative came to nothing, which was one of the reasons why my research team and I decided to complete the survey of the entire Isles ourselves. You might be surprised, as I was when I first heard of the Academy, that it still retains the Royal prefix, but it is one of the institutions that has survived the 1921 partition of Ireland. It was founded in 1785 and soon became the premier learned society for the study of Irish civilization. It is in many respects the Irish equivalent of the elite academic societies in the rest of Britain, like the Royal Society and the British Academy in London and the Royal Society of Edinburgh and, also in Edinburgh, the Royal

Scottish Academy. One of the enjoyable aspects of visiting these places is that they are almost always housed in grand Georgian terraces. It is always a treat for the eyes to attend meetings in such sumptuous surroundings, with the olympians of academe, Newton, Darwin and co., peering down from their portraits high up on the walls.

Unlike its British counterparts, the Irish Academy is not restricted to particular fields of endeavour. While in England, for example, the Royal Society deals with the sciences and the British Academy covers the humanities: literature, history, philosophy and so on, the Royal Irish Academy does not restrict itself, embracing both the sciences and the arts under one roof. This breadth made the Academy the natural home of a comprehensive survey of Ireland which would integrate all the diverse strands of science, history, language and archaeology. This irresistible combination, together with some fundraising, led to a substantial amount of money being made available to the Academy from the National Millennium Committee of Ireland. Invitations to bid for money from the fund went out to all the Irish universities and I found myself on a plane to Dublin to help to judge the applications.

In the elegant surroundings of the Academy's headquarters in Dawson Street, the hopefuls presented their proposals in the form of short talks. Naturally enough, when a new pot of money un-expectedly becomes available, people build their bids around their existing expertise. The aim is to persuade the judges that what they are already doing will, with a bit more money, produce an essential and indispensable contribution to the project. Our job, as judges, was to weigh up these diverse claims and to recommend where we thought the money would best be spent. You'll not be surprised to hear that I didn't need much persuading that a survey of mitochondrial and Y-chromosome DNA would be not only relevant but completely essential.

Fortunately, all the other judges felt the same way. Which is how Dan Bradley and his team from Trinity College Dublin got the green light – and the funds – to take charge of that central aspect of the project. Dan had pioneered the use of DNA to find out how and when farm animals, in particular cattle, had been domesticated. We had worked together a little on this when Jill Bailey, one of the research students in my team, had been working on retrieving DNA from the bones of the extinct ancestor of domesticated cattle, the fearsome aurochs. After getting her degree in Oxford, Jill had spent a year in Dan's lab in Dublin and I had been over a couple of times to give talks and be an examiner for Dan's graduate students. All of which is entirely irrelevant, except that it meant that a highly experienced and competent geneticist, whom I knew and liked, would be covering the same genetic ground in Ireland as I was already starting to do in the rest of the Isles. Which in turn meant that I could concentrate on Scotland, Wales and England, knowing that Dan's lab would produce compatible genetic data from Ireland that could be integrated with the results of our Irish customers at Oxford Ancestors and, eventually, all the data from the rest of the Isles. Which is precisely what did happen and it is from this grand coalition of data that we begin our tour.

On my imaginary map, I moved all the Irish gene-coins across the Irish Sea and began to distribute them according to their geographic origins. Each one of these was the end point of a journey – the journey of a line of ancestors stretching through maternal and paternal threads way back into the deep past. We know, from the archaeological records, that every one of these ancestral journeys must have ended in Ireland within the last

9,000 years when the first Irish built their timber-framed houses on the banks of the River Bann at Mount Sandel. Before that, as we know, Ireland had been uninhabited since the Ice Age. Could it be that some of these DNA fragments from today's Irish men and women have actually been there all that time? How would we know?

I began with the maternal DNA. I knew, from the identity of their clan mothers, when and where all these journeys had begun. And from the locations on the map of Ireland, I also knew where these journeys had ended. Well, not ended, because many of these genes will go on travelling the world for millennia to come. To be more accurate, I knew where these immortal time travellers had reached by the late twentieth century.

The longest journeys, in both time and distance, had been travelled by Ursula's descendants. Ursula herself had lived in Greece about 45,000 years ago, at the start of the Upper Palaeolithic, and had shared the land with the far more ancient Neanderthals. We knew this date from counting up all the mutational changes that had happened among her descendants and dividing this figure by the mutation rate. As we have seen, among the seven principal clans in Europe there were far more changes in the descendants of Ursula than in any of the others. That simple fact meant the clan of Ursula was the oldest of the seven. We arrived at the age of the clan, and thus the time in the past when Ursula herself lived, by factoring in a mutation rate of one change in every 20,000 years. From thousands of DNA samples from all over Europe we knew that, on average, Ursulans have 2.25 mutations in their mitochondrial DNA compared to the DNA of Ursula herself. That puts the age of the clan at 2.25 × 20,000 = 45,000 years.

We arrived at Ursula's own location in Greece by looking to see where the clan today is both the most frequent and at its most diverse. So long as there was good archaeological or climatic evidence that the location was inhabited, or at least habitable, at the time, then that was where we placed the matriarch. I realize from many letters that it is a frequent and understandable mis-understanding (if there can be such a thing) that we have located the skeleton of Ursula and the other matriarchs and then worked out how long ago they lived from carbon-dating. But this is not so; it is all accomplished by reconstructions.

From the DNA fragments now displayed on the map of Ireland, I could see that almost exactly 10 per cent of Irish men and women are the direct maternal descendants of Ursula, each carrying her DNA modified by the occasional mutation. Converting that proportion of 10 per cent to actual numbers of people means that of a total population of 5.7 million there are roughly 570,000 Ursulans in all of Ireland. Altogether, counting Dan's published data, we have the mDNA sequences of 91 Ursulans from a total of 921 Irish samples. However, because the clan is so old and there has been such a long time for mutations to accumulate, we found only three people who have Ursula's sequence unmodified by genetic change. As all three are customers of Oxford Ancestors, I can trace their present where-abouts and none of them lives in Ireland! One lives in Hampshire in southern England, another in London and the third in New York. It is their grandmothers and great-grandmothers who lived in Ireland and who then joined the stream of emigrants in the nineteenth and early twentieth century. But their DNA did pause in Ireland before it resumed its journey around the globe.

I shall refer to these three as 'pure' Ursulans, while, I hope, avoiding the implication that mutations in the others have some-how sullied the pristine genetic heritage of Ursula herself. I hope

I am not accused of implying that their DNA is, in contrast, somehow impure, which is complete nonsense of course. But it has been modified. Of the other Irish Ursulans, some have one change on the 'pure' Ursulan background – and I call them first-generation Ursulans. There are 22 of them. By the same token, there are 23 second-generation, 26 third-generation, 10 fourth-generation, 5 fifth-generation, one sixth-generation and one seventh-generation Ursulan. The third-generation Ursulans, with three mutations compared to Ursula herself, are the most frequent, closely followed by Ursulans of the second 'mutational' generation with two changes compared to the original. The numbers in higher generations tail off slowly until we reach the Irish Ursulan record-holder, a doctor now living in Chichester, in Sussex, with seven changes since Ursula. I had better stress once again that the 'pure' Ursulans and all the others up to and including the record-holder are separated from Ursula by the same 45,000 years and, roughly, the same number of actual maternal generations. It is chance alone that has left the three 'pure' lines untouched by mutations while the good doctor's has been hit seven times.

We are in the most technical part of the book and I beg for your indulgence to explain a very important point. From the numbers of 'pure' and first-, second- and higher-generation Ursulans, we can work out the average number of mutations over all the Irish Ursulans. It comes to fractionally over 2.5. If we now multiply this by the mutation rate of one change in every 20,000 years, it comes to just over 50,000 years, which is older than Ursula!

Actually, this is not that far from the 45,000-year date for Ursula herself and well within the mathematical error of the estimate. But it is an awful lot longer than the 9,000 years we know that people have been in Ireland. How can we explain this apparent discrepancy? We have 50,000 years' worth of accumulated mutation in an island which we know has only been inhabited for 9,000 years. It has to mean that most of the mutations in the Irish Ursulans must have already occurred *before* and not *after* they arrived in Ireland. We cannot just use the 50,000-year genetic date and say that is when Ireland was first inhabited.

My colleague Martin Richards and I got into a lot of trouble when we first used a superficially similar argument to back up our controversial proposition that the ancestors of most Europeans were Palaeolithic hunter-gatherers who had arrived a long time before the Neolithic farmers. We said that there was far too much accumulated mutation in all the major European clans, save Jasmine, to have developed in the 10,000 years or so since farming had been invented in the Middle East, and therefore these clans were Palaeolithic in origin. The counter-argument was vigorously expressed in terms of a Martian metaphor. Suppose a representative selection of Europeans had been transported to Mars and then, a few years later, had their DNA sampled and analysed – presumably by Venusians who didn't know about the landing. They had then done the calculations and showed that, according to the amount of accumulated genetic differences, the Martians had been there for tens of thousands of years, whereas we know they had arrived only a few years earlier. The flaw in the Venusians' argument – and by implication in ours – was that they had assumed that all the mutations had accumulated *after* the earthlings arrived on Mars, when, in fact, they had all occurred *before* they set off from Earth.

Although Martin and I, for a number of reasons, did not think this was a good analogy for what we had actually done, we none the less set out to try to prove that the mutations in Europeans (we are back on Earth now) had accumulated in Europe and had not been imported from the Middle East by Neolithic farmers. Martin particularly, helped by our new recruit, theoretical physicist and mathematician Vincent Macaulay, spent three long years doing this.

To cut a very long story short, they scoured the Middle East for as many DNA samples as they could find, then searched these for matches to the DNA from Europe. The point was to find out how many of the mutations in Europe were genuinely European and how many had already happened in the Middle East. Basically, if a match was found, and if we could be certain that it had actually originated in the Middle East, rather than being carried back there from Europe by some sort of reverse migration, we subtracted it from our tally of 'European' mutations and did the time calculations once again. It was an exhaustive, and exhausting, analysis which in the end gave us a set of dates for the settlement of Europe that we could all rely on. Fortunately, they were not so very different from our original ones and did not reverse our conclusion that most Europeans had hunter-gatherer ancestors.

Turning back once more to the Irish Ursulans, could I do the same sort of thing to work out which mutations were Irish 'originals' and which were imported? If I could, then the genetic dates would mean something. This I did by checking each of the Irish Ursulans' DNA sequences against every other sequence that I knew about from all over, first Europe, and then the world. I was looking to see how many of them had also been seen outside Ireland. Although my computer helped a great deal in sorting all the results so that identical sequences appeared on consecutive lines on the screen, I also checked the list of Irish Ursulans one by one. One Irish Ursulan, a lady from Donegal, for example, had only one matching sequence and that was a man from the Czech

Republic. It is very unlikely we will ever know the precise tracks that trace the wanderings of the ancestors of these two genetic relatives back to the woman whose DNA they both share. But it is in just these tracks, like footprints in the sands of time, that we can read the signals from the past.

I found, at the end, that out of the 91 Irish Ursulans, 68 had matching sequences elsewhere. Only 23 were unique to Ireland. In many cases I could find their immediate predecessors, in a genetic sense, within Ireland. So a fourth-generation Ursulan sequence, for example, would usually have a third-generation sequence nearby. In these cases I thought it was reasonable to regard that fourth mutation as having happened in Ireland. Counting up these home-grown mutations and factoring in the mutation rate as usual gave me a corrected date for the clan in Ireland of a little over 7,000 years – 7,300 to be precise. This was much more reasonable than the 50,000 years which counted all the Ursulan mutations as if they had all happened in Ireland.

Genetic dates, like the 7,300 years for the Irish Ursulans, are not very accurate. They are estimates. We find this concept particularly difficult to grasp because we are accustomed to dates being very precise. The 7,300-year date for the arrival of the Irish Ursulans is an estimate. The date could vary a thousand years either way and still fall within the scope of the estimate. Forgetting the inaccuracies for the moment, what does this date mean? It is an estimate for the length of time it would have taken for all the Ursulan mutations to have accumulated within Ireland. If the ancestors of all 91 Irish Ursulans had arrived at the same time and their mDNA mutations had accumulated since then, the genetic estimate for their arrival would have been 7,300 years ago. Of course, it is very unlikely indeed that they all arrived at once. Some would have come more recently, but in that case, to achieve the average figure, others must have arrived *more than* 7,300 years ago to balance out the more recent arrivals.

I went through the same procedure with all the other Irish maternal clans, checking to see in each one how many looked as though they had mutated to their final form in Ireland and not elsewhere. From there I calculated the clan arrival times in the same way as I had for the Ursulans. All of them came out between 7,500 and 4,500 years ago. Ursula was still the oldest clan in Ireland and, in common with the rest of Europe, Jasmine was the youngest. It was Jasmine's clan that Martin Richards and I had linked to the arrival of Neolithic farmers in Europe from the Middle East. The others clustered around the 5,000–6,000-year average. Even bearing in mind the approximate nature of these genetic dates for the settlement of Ireland by the various maternal clans, they are all way before the time, around 200 BC, when the Iron Age Celts were supposed to have arrived. It was beginning to look as if the ancestors of today's Irish had been there for a lot longer than anybody thought.

But within Ireland, when I looked at the maternal clans in the different provinces of Ulster, Leinster, Munster and Connacht, there was very little noticeable difference between any of them, though the numbers in each were too low to be sure of statistical significance.

If women had been in Ireland for a very long time, what about the men? Just as mitochondrial DNA traces our maternal ancestry, so the Y-chromosome follows paternal genealogies. Like mitochondrial DNA, Y-chromosomes also experience random mutations over the course of time. The precise nature of the mutations might be different between mDNA and Y-chromosomes, as we shall see, but the principles are the same. The accumulation of mutations along paternal genealogies over a very long time means that there are now tens of thousands of slightly different Y-chromosomes which can be distinguished by genetic tests. If two men have the same Y-chromosome fingerprint, then they have usually inherited it from a common patrilineal ancestor. That's exactly the same principle as saying that if two people have the same mitochondrial DNA sequence they have inherited it from a common maternal ancestor. Although Y-chromosomes are quite different from mitochondrial DNA in the way they change genetically, that doesn't matter so much when it comes to interpreting the signals they are bringing us from the past.

Just as each of us belongs to one of a small number of maternal clans, so men can be assigned to a paternal clan by the genetic characteristics of their Y-chromosome. From research done throughout the world over the past decade, Y-chromosomes can now be separated into twenty-one paternal clans, eight of which occur in Europe. Of these eight clans only five occur in the Isles to any appreciable extent. Following the tradition of the maternal clans, I have given them names. They are the clans of Oisin, Wodan, Sigurd, Eshu and Re. Like the maternal clans, each founded by a matriarch, the paternal clans must, by the same logical inevitability, also have been started by a single man – the clan father or patriarch. Every man within a clan is a direct paternal descendant of the clan father and has inherited the patriarch's Y-chromosome, modified by mutations over the intervening millennia.

The different paternal clans are told apart by single DNA changes, just like the mutations in mitochondrial DNA. However, these Y-chromosome sequence changes occur far more slowly than they do in mDNA. There is usually only a single DNA sequence mutation between the Y-chromosomes of one clan and another, even though they have been separated for tens of thousands of years. Luckily for us, the Y-chromosome also experiences a second, much swifter, type of mutation that can split a paternal clan into hundreds, if not thousands of separate paternal lineages. These fast mutations happen, like all DNA changes, when cells divide and there is an error in the usually immaculate copying mechanism.

Along the Y-chromosome are patches of DNA sequence that, when looked at closely, consist of reiterated blocks of short sequences. Treating DNA sequences like a word, these are the genetic equivalent of a bad stammer. It is as if an otherwise smooth read-through just gets stuck. Take the four DNA letters TAGA, an outwardly unremarkable sequence. For some reason, TAGA tends to trip up the DNA copying mechanisms on parts of the Y-chromosome where it is repeated a number of times. Cells can handle a few repeats. The double reiteration TAGATAGA causes no difficulties. Even ten repeats, one after the other, is manageable. But after that the stammering begins. After twelve repeats cells find the blocks of TAGA very difficult to copy accurately and make mistakes much more often than they normally would with regular sequences. What they do wrong is to add an extra TAGA, or forget to copy one. So a Y-chromosome with, let's say, fourteen TAGA repeats mutates into one with fifteen repeats. Because the cell has such trouble with copying this type of stammering sequence accurately, the rate at which these mutations occur is hundreds of times faster than the regular type of spelling-change mutation, where, for example, a C changes to a T. It is an even faster rate than mDNA, with its comparatively lax error-checking mechanisms.

There are dozens of places along the Y-chromosome where these tricky stammering segments are to be found and they can be used in combination to define tens of thousands of different

Y-chromosomes. Because Y-chromosomes, alone among the nuclear chromosomes, are not shuffled at each generation, the combinations can persist for a very long time, changing only when another mutation occurs at one of them. In this respect, the Y-chromosomes can be interpreted just like the mDNA with first-, second- and third-generation mutations changing the Y-chromosome fingerprint of the 'pure' patriarch. And, just like mDNA, the number of mutations can be added up to get an idea, again only approximate, of time passing. In reality, because these mutations happen so quickly, in relative terms, it is hard to know what the clan patriarch's Y-chromosome fingerprint actually was. But as we shall see, this is not really a problem when we look in detail at the Y-chromosome of Ireland.

In my mind's eye, I collect up the maternal gene-coins from the imaginary map of Ireland, move them to one side and begin to distribute the paternal Y-chromosome equivalents in their place. As soon as I look at the Irish pile, one thing stands out. The vast majority of Irish Y-chromosomes are members of just one clan, the clan of Oisin. It is precisely because of its predominance in Ireland that I gave the clan this name. Oisin was the son of the hero of another of the Irish mythical cycles, Finn mac Cumhaill, sometimes transcribed as Finn mac Cool. Finn is the leader of the Fianna, a band of warriors chosen only after an appropriately gruelling selection process. His son Oisin, meaning Little Deer, is bewitched by Niav of the Golden Hair, the daughter of the Underworld king who reigns over Tir na n'Og, the World of the Forever Young. Oisin goes to this other world to be with Niav and spends his life writing poetry and songs. Eventually he becomes homesick and is eager to visit his own world once more. Niav warns against this, but Oisin is adamant and sets out, though he promises to heed her warning not to set foot on Irish soil. He plans to avoid disobeying Niav's instructions by riding everywhere on a horse and so not touching the soil. In an extreme version of *Back to the Future*, Oisin realizes when he returns to Ireland that 300 years have passed while he was relaxing in Tir na n'Og. The shock of this discovery makes him fall from his horse and, as soon as he touches the ground, he instantly ages 300 years and crumbles into dust.

The Y-chromosome might be decaying fast, and well on its way to sharing Oisin's fate, but for the moment his clan's Y-chromosome is doing extremely well in Ireland. Almost 80 per cent of Irish Y-chromosomes belong to the clan of Oisin. Within Ireland there was very little difference to be seen in the geographical distribution of the maternal clans in different parts of the island. However, with Y-chromosomes there certainly is. Dividing Ireland into the four ancient provinces, each roughly occupying a quadrant of the Irish rectangle, the differences between them are very striking indeed. In the south-east quadrant of Leinster, 73 per cent of Y-chromosomes are in the clan of Oisin. In Ulster, in the north-east, this rises to 81 per cent. The clan reaches an even higher frequency in the province of Munster, in the south-west, where 95 per cent of men are in the clan. However, in Connacht, occupying the north-west quadrant, the proportion of Y-chromosomes in the clan of Oisin reaches an astonishing 98 per cent. This was one of the first results from Dan Bradley's genetic survey of Ireland, which he undertook in this instance with graduate student Emmeline Hill. But, rather than just leaving it at that, Dan took their analysis one stage further to look for an explanation. Recalling the twelfth-century Anglo-Norman invasion of Ireland, it occurred to them that one reason for the difference in Oisin Y-chromosome frequency between the provinces might have something to do with this invasion and the subsequent occupation. The invasion began in Leinster, in the south-east, and that was where the Oisin clan was in its lowest proportion.

The vehicle for testing this idea was to use surnames which, like Y-chromosomes, are also passed down the paternal line. There have been inherited surnames in Ireland for as long as any-where in Europe. They were first adopted in about AD 950, a good 200 years earlier than in England. The Gaelic origin of many Irish surnames is evident from the prefix 'Mc' or 'O', meaning 'son of', as in McCarthy or O'Neill, but there are plenty more whose Gaelic origins required a little research. Fortunately, the enormous interest in genealogy over the last hundred years has led to the compilation of comprehensive surname dictionaries where the origin of almost any name can be found. Sure enough, when Y-chromosomes were compared to surname origins, Gaelic or Anglo-Norman, the correspondence was clear. Even in Leinster, the proportion of Oisin clan members was higher among the men with Gaelic names than among the men whose surname could be traced to Anglo-Norman origins. You will recall that, in his survey of blood groups, Professor Dawson explains the higher frequency of blood group A in Leinster by the Anglo-Norman invasion.

The comparison with the mitochondrial results is striking. All of the seven major maternal European clans, and most of the minor ones, were to be found in Ireland and there was not much difference in their proportions in the four provinces. Very obviously, in Ireland anyway, the version of history told by men and women was not the same. To explore this further, I want to look in more detail at the Irish Y-chromosomes, in particular the detail among the members of the clan of Oisin. A Y-chromosome fingerprint, or signature, consists of a set of ten numbers. Each of these is the number of stammering DNA repeats at the ten places on the Y-chromosome that we test. If there are 10 repeats at the first marker and 22 at the second one, the fingerprint starts off as 10–22. If there are 13 repeats at the third marker, the fingerprint continues as 10–22–13 and so on. When I am checking through 10 marker signatures from the DNA analyser, if there are Irish men among the batch it isn't long before I find this particular signature: 11–24–13–13–12–14–12–12–10–16. It is very familiar indeed. This is the quintessential Oisin Y-chromosome and huge numbers of Irish men carry it. I am not the only person to have noticed this particular Y-chromosome combination. Dan Bradley certainly knows about it, and it has also been spotted by Jim Wilson, who worked for a time at University College London. Jim is a native of the Orkney Islands, just off the north coast of Scotland, and he had noticed this same combination among his fellow Orcadians. It also cropped up, interestingly, in surveys of Y-chromosomes among the Basques of north-eastern Spain and among the people of Galicia in the north-west of Spain. Its oceanic affinities led it to be christened, not Poseidon or Neptune, but the far more prosaic Atlantic Modal Haplotype, or AMH for short. I prefer to call it the 'Atlantis' chromosome. In the Isles it is by far the commonest Y-chromosome signature within the clan of Oisin, or any other for that matter.

Among the Oisin clan in Ireland, it certainly isn't the only Y-chromosome fingerprint, but most of the others can be linked to it by one or two mutations. This gives us an opportunity to get an Irish date for the Oisin clan, rather as we did with the Ursulans and other maternal clans. The mutation rate of these Y-chromosome fingerprints is roughly one change per 1,500 years, much faster than the mDNA rate of one for every 20,000 years. By following exactly the same procedure for the Oisin clan as we did for the first Irish Ursulan calculation, we get a date of 4,200 years. This lies well within the time frame of human settlement in Ireland and, since it is still prey to the wide approximations of genetic dates, and thus quite close to the 5–6,000-year estimate for mDNA, it would be tempting to imagine we have solved the origins of the Irish.

However, we have done nothing of the sort, because we have overlooked the Martian factor. Remember that the original Irish Ursulan date was 50,000 years ago – older than Ursula herself and far too early to be a plausible date for the settlement of Ireland – assuming that all the mutations among Irish Ursulans had happened after the first ones reached Ireland. To get round this we had to decide which of the mutations happened on Irish soil, and which had already occurred before the clan reached Ireland. When we did this it made a huge difference to the date, bringing it forward to the much more plausible 7,300 years ago. Applying the same correction to the Irish Y-chromosome data brought the date forward to just 1,200 years in the past. This is far too recent to be plausible, even given the approximations involved. Something else must be going on. And it was, but it took me a while to realize the explanation. And it didn't happen until I had done a lot more work. Not in Ireland, but in Scotland. Only then was I able to make sense of the strange behaviour of the Irish Y-chromosomes.

Nevertheless, we have made a good start. In Ireland, the maternal lineages are diverse and very old, while the Y-chromosomes are unexpectedly homogeneous, and at first glance look comparatively young. We have seen a difference between different regions of the island, a difference that may be an echo of the Anglo-Norman invasion of Ireland beginning in the twelfth century. We have seen some evidence of a genetic link between

Ireland and Spain along the Atlantic fringe of Europe, which archaeologists are now beginning to realize was a much busier seaway than was once thought. What we don't yet know is how the Irish results will fit with the rest of the Isles, and to begin to do that we shall travel across the shallow sea to Scotland.

10

SCOTLAND

It is barely 12 miles across the sea from Fair Head on the north-western tip of Ulster to the cliffs of the Mull of Kintyre, rising above the waves of the North Channel. Scotland is bounded on three sides by the sea: by the wild Atlantic to the west and north and by the temperamental North Sea beyond the eastern coastline. Across its historically fluctuating southern land boundary lies England, at different times enemy and friend, but never indifferent neighbour. The western sea boundary is fringed with several large, inhabited islands and hundreds of small ones deprived of inhabitants. Off the north coast lie the Orkney Islands, and 60 miles further to the north-east and halfway to Norway are the Shetlands. The total land area, including the islands, is just over 30,400 square miles, only slightly smaller than Ireland. Mountains dominate the mainland, with the rugged Scottish Highlands reaching to over 1,300 metres. Ben Macdui (1,309 metres), highest of the Cairngorms in the north-east, and Ben Nevis (1,344 metres) in the west are the highest mountains in the whole of the Isles.

The mountains continue all the way to the northern coast of Scotland, especially on the west side, where millions of years of erosion, compounded by the gouging action of the glaciers, which covered the whole of Scotland in the last Ice Age, have left a dramatic landscape. In the far north-west, Old Red Sandstone peaks like Suilven and Stac Pollaidh stand isolated above feature-less country of bog and lochan. In the extreme north, the mountains relent, leaving a fertile coastal strip where the thin, acidic soil of the Highlands is invigorated by calcium-rich lime-stones and sandstones.

The effect of limestone, wherever it occurs, is always dramatic. It neutralizes the otherwise acidic soils and in so doing transforms the colour of the landscape from a yellow-brown to a vivid green. In the Highlands and the Hebrides, the occasional limestone out-crops are marked out by the rich growth of grass and wild flowers. But nowhere is the effect of neutralizing the soil more noticeable or more delightful than in the Western Isles, the long chains of islands that protect the mainland from the full force of the Atlantic. On the western edge of these islands are some of the most beautiful beaches in the world. Brilliant white in the sun-light and lapped by turquoise, translucent seas, they are not made of the usual sands to be found on the crowded holiday beaches of southern England. The white beaches of the Western Isles are composed of the pulverized shells of countless billions of sea creatures that have been ground to a coarse powder by the pound-ing waves of the Atlantic. The wind, which for 300 days out of 365 roars in from the ocean, has blown the shell sand inland for a mile or two. And there it works its magic on the soil, neutralizing the acid and supplying essential phosphates that are otherwise entirely lacking. The result is the *machair*, a thin strip of meadows and grassland which, so long as the sheep don't get there first, is full of wild flowers – purple orchids by the hundreds, blue harebells and the purple and yellow flowers of heartsease, the wild pansy. A couple of miles further inland, beyond the reach of the wind-blown shell sand, the moss and dark rushes are back, signalling the return of the acid lands.

The white beaches are also spread along the north coast, but there they are not needed to help the soil. The older gneisses and schists of the Highlands, among the oldest rocks in the world, are replaced by alkaline sandstone. Green grass grows far inland in Caithness at the extreme north-east tip of the mainland, and is rich enough to support large herds of sleek black cattle. The fertility of the sandstone soil is even more remarkable in the Orkneys, now a few miles from the Caithness coast but joined to it until 7,000 years ago. On the east coast, there is good low-level farmland around the Moray Firth near Inverness and inland of Aberdeen, at the eastern edge of the Cairngorms. One deep geological fault divides the Highlands along the Great Glen, running between Inverness and Fort William. Another fault line runs between Stonehaven, on the east coast just below Aberdeen, and Loch Lomond to the north of Glasgow. This southern fault line separates the Highlands from the rich farmland of the Central Lowlands, which is also the location of the major cities of Dundee, Stirling, Glasgow and Edinburgh. Most of the 5.2 million Scots live in this Central Belt, a great many having moved there from the Highlands. Further south the ground rises again to form the hills of the Southern Uplands. Lower and less rugged than the Highlands, these hills have been eroded by glaciers into smooth-topped plateaux separated by narrow, flat-bottomed valleys. Beyond the hills, the valleys open out into the rolling farmland that surrounds the River Tweed, which flows into the

North Sea at Berwick on the east coast. On the west side of the Southern Uplands, the hills give way to the Galloway peninsula and the flat lands bordering the Solway Firth.

Since the whole of Scotland was under thick ice until the end of the Ice Age and again during the cold snap of the Younger Dryas, it isn't surprising that no evidence, yet, has been found in Scotland of Palaeolithic settlements such as remain in the Cheddar Caves in south-west England. The first signs of human occupation are not found until well after the cold snap and, as in Ireland, these are Mesolithic settlements at or near the coast. The earliest dated site is at Cramond, on the southern shore of the Firth of Forth, only 3 miles from the centre of Edinburgh. It is a picturesque spot, with a small terrace of old houses on one bank of the River Almond, where it flows into the Firth. Swans and ducks bob around in the quiet tree-lined bay and, when I visited on a crisp sunny day in November, I could not have imagined a better spot for a bit of hunter-gathering. A seashore for shellfish and wading birds, a medium-size freshwater river for salmon. All that would have been missing was the cappuccino that was steaming on the table in front of me. The Cramond remnants, a few microliths and the bony evidence of past meals, are dated to about 10,000 years ago. There are no signs of permanent settlement at Cramond, no post-holes as at Mount Sandel in Ireland, so it was probably just one of many places where the small bands of humans used to camp for a while as they moved around the country in search of food.

The seasonal movements of the Mesolithic hunter-gatherers from one site to another are nowhere better illustrated than on the island of Oronsay, off the opposite coast of Scotland from Cramond. Oronsay is a small island, roughly triangular in shape and each side only 3 kilometres long. Despite its small size, no less than five Mesolithic shell middens have been discovered, each containing vast numbers of mollusc shells. Limpets, winkles, whelks, oysters and scallops were all on the menu. Curiously shaped implements, made from the antlers of red deer, have also been found. Their use is immediately obvious when you watch the staff at work in an oyster bar. They are shaped exactly like the knives which, inserted between the two shells of an oyster, then twisted, open it to reveal the silver-grey flesh inside. It is a sight to behold – as it must also have been for the children of Oronsay, 8,000 years ago, for that is the date of the Oronsay midden.

The seasonality of the Oronsay middens has been discovered in a very curious way. As well as huge numbers of mollusc shells, the middens also contain the bones of saithe, a relative of the haddock that is still plentiful in the waters off the west coast of Scotland. The saithe grows rapidly in its first years of life and the age of a fish can be worked out from, of all things, the length of the ear bone or *otolith*. Otoliths within the same midden tend to be about the same length but there is a big difference in average otolith size between one midden and the next. The conclusion is that the middens marked different seasonal camps where the fish caught were at different stages of their development. What we do not know is if Oronsay was a permanent home or, like Cramond, another seasonal camp, occupied at the same time each year to take full advantage of the harvest of the sea.

Oronsay and its close neighbour Colonsay lie about 15 miles from the larger islands of Islay and Jura, themselves 10 miles or so from the long finger of the Kintyre peninsula. Clearly, the Mesolithic hunter-gatherers – an epithet to which we must surely add fishing – were well used to making these quite substantial sea crossings between the islands and the mainland. No boats remain, destroyed by millennia of decay, but they were probably made from animal skins stretched across a framework of hazel branches. They would have resembled the coracle, still, just about, used for fishing in the rivers of west Wales, and the more substantial curraghs of the west of Ireland. Whatever they used, these boats were perfectly good enough for coastal work and island hopping.

The sea has never been a barrier to the people of the Atlantic. It was their highway, just as the Pacific was to the Polynesians. There are confirmed Mesolithic sites on many of the islands lying off the west coast of Scotland, and where no evidence has yet been found there is a feeling among archaeologists that, with more field work, every island will be shown to have been occupied, if only for one part of the year. There is even indirect evidence, in the form of unusual patterns of soil erosion, that the Mesolithics reached Shetland, which would have involved a voyage on the open sea of 60 miles from Orkney, the nearest point. Valuable materials were also transported over long distances by sea. Flint is unknown in Scotland and other stones were used for making tools. Bloodstone quarried from the Isle of Rum, where there is a very early Mesolithic settlement, has been found in many sites around the west coast. The Mesolithic was a time of plenty for those bands who lived on the coasts of Scotland and Ireland. There was ample food within easy reach, both in the sea and in the dense woodland that lay behind the shoreline. It certainly wasn't crowded. One recent estimate puts the total population of the whole of the Isles during the Mesolithic at less than 5,000.

There is one tantalizing fragment of evidence – a grain of wheat pollen from the Isle of Arran in the Firth of Clyde – that the Mesolithics were already experimenting with growing their own plant food, well before the arrival of agriculture proper. However, it is only with the arrival of farming that the whole way of life begins to change. Curiously enough, despite the major effect this transition from the Mesolithic lifestyle of thinly dispersed hunter-gatherers to full-blown farming must have had on the early inhabitants of Scotland, there is a distinct lack of material evidence from the early stages. Part of the reason is probably the later growth of thick layers of moss which have buried early field systems. In Ireland a whole patchwork of fields has been discovered at Ceidi, near Ballycastle, County Mayo in the north-west, lying under several feet of peat and visible only when this layer was cleared away. Even the megaliths suffered from the accumulation of moss and peat. The stone circle at Callanish on Lewis, where the stones reach nearly 5 metres in height, had been almost swallowed up by the peat before it was excavated in the nineteenth century. Only the tips of the tallest stones protruded above the peat.

Covering of a different kind obscured what, in my opinion, is the most remarkable archaeological site in the whole of the Isles. The settlement at Skara Brae in Orkney does not have the grandeur of Callanish or Stonehenge. It is altogether more domestic. Following a violent storm in the 1850s, the sand dunes which back on to the beach in the Bay of Skaill, on the west coast of the largest island, were stripped back to reveal the walls of houses. Unlike today, when such a discovery would precipitate an immediate excavation, nothing much was done either to excavate or even to protect the site until the early years of the twentieth century. Hidden beneath the sand was a small group of interconnecting stone houses, each about 5 metres in diameter and complete with stone beds, stone dressers, even waterproofed stone basins sunk into the floor to keep live lobsters and to soften limpet flesh for fishing bait.

That Skara Brae is still standing and not strewn about the countryside has a lot to do with the remarkable rock found all over Orkney and Caithness. The sandstone comes in flat slabs, about 5–10 centimetres thick. Even without mortar, anything built with Orkney flagstones is not going to fall down. Ruined buildings, 100 years old, which are a not uncommon sight all over rural Scotland, are still standing. Their roof timbers have decayed and collapsed, but the walls of flagstone houses are as solid as ever. Metre-square flagstones, split even thinner, are even used as roof tiles or stuck upright in the ground as fencing.

The charm of Skara Brae is in its ordinariness. I have to admit that, though I enjoy standing in awe amidst the great monuments from the past, I feel strangely detached from them. But at Skara Brae I really *can* imagine people living there, coming in from the wind to the warm, snug interiors, recounting, in whatever tongue, the events of the day. The beach at Skaill just next to Skara Brae is strewn with broken flagstones and when I was there, during the school summer holidays, families were playing on the beach. But instead of building sandcastles – and there is plenty of good sand – the children were constructing their own miniature stone circles. These rocks are just asking to be stood upright, and that's exactly what has happened all over Orkney. The Ring of Brodgar, about 5 miles inland from Skara Brae, was originally a circle of sixty stones 7 metres high and 100 metres across. Twenty-one remain in position. A mile in one direction is the stone circle of Bookan, while the same distance in the other direction is another, Stenness, and half a mile further lies the astonishing passage tomb of Maes Howe. Like the tomb at Newgrange on the Boyne, Maes Howe is aligned so that the sun shines along the low passage at the winter solstice and floods the inner chamber with light. Once again, the wonderful building quality of the rock makes Maes Howe appear much younger than its 5,000 years, the stone slabs neatly laid and corbelled at the top to form a roof. These are only the major structures. All around are burial mounds, many not yet excavated, single standing stones and other remnants of a vibrant ritual past.

The sheer scale of the Orkney megalithic monuments, and the equivalents in Ireland and all along the Atlantic coast, is a testament to the economic effects of agriculture. However like us they may have been, the Mesolithic hunter-gatherers who first settled in Scotland simply could not have assembled the manpower to build these great monuments. There just were not enough of them. The easy life on the shoreline could support only a few thousand people. It was the coming of agriculture to Scotland, beginning about 6,000 years ago, that boosted the population so that, only a few centuries later, there was enough manpower to construct these vast monuments. But did this evidently greatly increased population mean the immigration of large numbers of people, or did the original Mesolithic inhabitants adapt and proliferate? Were the descendants of the fishermen of Cramond and Oronsay replaced, or at least overlaid, by new arrivals? There is no firm archaeological evidence either way, and it is one of the principal questions to ask of the genetic evidence.

There is, however, on Orkney, ample evidence that, whether or not the inhabitants of Skara Brae and the builders of Maes Howe were descended from the original stock, they were not by any means the last people to take an interest in this green and fertile land. In the centre of Kirkwall, the capital of Orkney, stands the magnificent medieval cathedral of St Magnus, built in the mid-twelfth century. It is as impressive a piece of late Norman architecture as anything in England. But Norman it is not. This is a Viking cathedral, started in 1137 by Rognvald, Earl of Orkney. Vikings began to arrive in Orkney, and in Shetland to the north, at the end of the eighth century. There is no exact date, but this coincides with the first of the Viking raids in England on the undefended monastery of Lindisfarne off the Northumbrian coast of England. The date of that raid is known very precisely. It took place on 8 June 793. The raiders carried off the rich monastery treasures and returned whence they came. That may have been Norway, but it is more likely that the base for the raid on Lindisfarne was Orkney.

Lindisfarne was the first of many raids. The next year a Viking fleet attacked Jarrow, down the coast from Lindisfarne. The following year, 795, the raids switched to the west coast of Scotland and St Columba's church on Iona was attacked. Iona suffered twice more, in 802 and 806. Enough was enough and the monastery was evacuated back to Kells in Ireland – well away from the coast. The Viking raids were only the first flurries in a campaign of invasion and settlement that dominated the Isles for the next 400 years. By the time Rognvald began to build his magnificent cathedral, the Vikings had long been in control of Orkney and Shetland. They had established bases in Ireland at Dublin, Waterford, Wexford and Limerick, though their hold there was always precarious and they never managed to get control of Ireland from the High Kings. But that was not for lack of trying. In the 830s a large Viking raiding fleet appeared regularly around the Irish coast and by the 840s they had built Dublin up as a major base for slaving and for attacking Britain. The Dublin Vikings took sides in the ceaseless wars between the feuding Irish kings, but made the bad mistake of joining the losing side and, in 902, quickly evacuated Dublin and retreated to the Isle of Man to escape the advancing army of their conquerors. But they were back in force by 907.

The Irish bases were at the end of a supply chain of men and weapons, based on Orkney and extending from the Hebrides to the Isle of Man and, on the mainland, to parts of Argyll. By AD 1000, Norse power had reached its peak and the Vikings were gradually forced back towards Orkney. They lost Limerick in 965, Dublin in 999 and were finally beaten at the battle of Clontarf in 1014. This famous battle is remembered not only for the Irish victory but also for the death of Brian Boru, King of Cashel and High King of Ireland. The Norsemen maintained their grip in Man and the Western Isles until well into the twelfth century, when they were driven out by Somerled, a Celtic hero we will hear more about. Eventually they lost Orkney and Shetland when the two island groups were annexed by James III of Scotland in 1468.

Although the initial Viking raids on Lindisfarne and the other coastal monasteries were motivated by material avarice, and the glory of returning home with conspicuous wealth, it was not long before the Vikings began to settle. Confined to a coastal strip of western Norway between the mountains and the sea, it is not hard to understand the attraction of the rich farmland of Orkney and the northern Scottish mainland. Shetland, too, which was much nearer home, was also attractive for settlement, though less fertile than Orkney. As well as being scarce, farmland in Norway was handed down to the eldest son. Younger sons had few prospects at home and the chance of getting land overseas was a great temptation. How extensive the Norse settlement of Scotland eventually became is one more question I hoped to answer using genetics. Certainly their cultural influence has been overwhelming. All place-names in Orkney and Shetland have a Norse origin, and *Norn*, a hybrid Scots/Norse dialect, was spoken there until the end of the eighteenth century. And yet positively identified Viking archaeological remains are few and far between in Scotland.

An exception is the site at Jarlshof on the southern tip of Shetland, where a small indigenous community was replaced by a series of Viking longhouses some time in the ninth century. These were impressive structures. Though only the base of the walls survives – the building stone is far more irregular in Shetland than in Orkney – the typical layout of a Norse long-house is easy to see. Built of stone with an earth core, the houses are typically 20 metres long by 5 across, with door spaces opposite each other in the long wall. There are two or possibly three houses at Jarlshof and their position on the site of early, circular, houses might indicate a violent settlement, although that cannot be taken for granted. Grouped longhouses, as at Jarlshof, are unusual; single isolated farmhouses are more the rule, giving no indication of whether the settlement was amicable or violent.

If the genetics proved that the Norse settlement in Scotland had been extensive, then who was it that had been replaced? This is the time to introduce the people who, above all others in the Isles, are surrounded by the greatest mystery – the Picts.

11

THE PICTS

On 15 July 1995 the final section of the Skye Bridge was lowered into place. Three months later, at 11.00 a.m. on 16 October, the Secretary of State for Scotland declared the bridge open and, for the first time since the Ice Age, the Isle of Skye was joined by solid connection to the mainland. The very next day the protests started. The target of the protesters was a combination of the very high toll, the loss of the ferry and the suspicion that the main financial backers, the Bank of America, were making far too much money. By 1.00 a.m. there was a queue of thirty-five cars all refusing to pay. Welcome to the spirited world of island protest. There followed years of active opposition, non-payment, even one imprisonment. Hilariously, those charged with refusing to pay the toll had to make the 140-mile round trip to the Sheriff Court at Dingwall, which once again meant crossing the bridge, where they of course refused to pay, thus incurring further criminal charges. The Skye Bridge toll became a *cause célèbre* until eventually the Scottish Executive bought the bridge and the tolls were scrapped on 21 December 2004.

Though Skye is firmly Gaelic, the protests were co-ordinated by one Robbie the Pict. Not just a sobriquet but a formal change of name. Born Brian Robertson, Robbie the Pict is a celebrated campaigner against all kinds of modern evil. He has been arrested over 300 times, refuses to pay road tax and has formed his own sovereign state on one acre of land in the north end of Skye. There are one or two other people who choose to link their name to this obscure ancient people. I am regularly contacted by one 'Nechtan the Pict', eager to enlist my help in recovering DNA from an allegedly royal Pictish body found in Perthshire. Clearly it means something to be thought of as a Pict. So who were they?

Although the Picts have been garlanded with an air of mystery, with book titles such as *The Puzzle of the Picts* that capture the imagination with hints of a lost people, the answer is almost bound to be more prosaic. The derivation of the name is *Picti*, the generic nickname the Romans gave to the indigenous inhabitants of the Isles. It was not just the northern tribes that were given this description. Any tribes the Romans encountered who either wore tattooes or adorned their bodies with wode earned the uncomplimentary nickname. *Picti*, literally 'the Painted People', is also from the same root as *Pretani*, the Gaelic term which, according to the Romans, the islanders used to describe themselves and from which, so some historians believe, the name Britain itself derives. As the Romans occupied more and more of the Isles and developed separate names for the tribes they conquered, the only peoples left with the original nickname were the tribes living in the far north. All tribes north of the Antonine Wall, which ran between the Clyde and the Forth, were automatically Picts.

The material remains of the Picts are extremely impressive, though nowhere numerous. About 200 carved symbol stones and rock inscriptions have survived, mainly in the north and east of the Scottish mainland and on Orkney. Even though they are for the most part badly weathered, the symbol stones reveal a mastery of naturalistic relief and abstract carvings. Most of the inscriptions and carvings date from the fourth to the seventh centuries AD, the later ones incorporating Christian symbols in the wake of St Columba's conversions. They are not close to any of the other contemporary styles to be found in the Isles, not Roman, Saxon or even Irish, adding further mystique to the Pictish enigma.

The Picts also left behind a collection of remarkable stone structures unlike any other in the Isles – the brochs. These take your breath away, especially when you realize that they were built over 2,000 years ago. Their form is similar wherever they are found. Round towers, tapering inwards at the top rather like a power station cooling tower, these huge stone buildings were once the largest structures in the whole of the Isles. Brochs typically enclose a central area 10–12 metres in diameter, with walls only slightly lower. They have double-skinned walls, held together by flat stones which form inner galleries within the walls. As well as providing storage space, these gaps, just like cavity walls in modern houses, would have insulated the interior and kept the heat in. This is easiest to see where there has been a partial collapse, such as at Dun Telve, near Glenelg on the mainland opposite Skye, or at Dun Carloway on the west coast of Lewis, a few miles north of the stone circle at Calanais. From gaps left in the inner walls, it looks as if the central area was fitted with wooden galleries, the whole structure resembling Shakespeare's Globe Theatre in miniature. The brochs were not roofed, so fires in the central living area could vent straight into the open air.

At first sight brochs appear to have been built for defence and they would certainly have been extremely difficult to attack successfully. The view from the ramparts would have given plenty of warning of any hostile approach and the blank, windowless external walls were impregnable. However, it is not at all certain whether brochs were built to withstand attack – or just to show off. Some archaeologists believe they are simply a natural evolution of the much smaller Pictish roundhouses typical of the region. Since there is no evidence of attacks, such as the reddening that discolours the stone of buildings that have been set alight, it is more likely that their impressive bulk was valued for just that purpose – to impress. The standard design, and the relatively short time over which the brochs were built, in the first and second centuries BC, suggests that there may have been mobile teams of masons and labourers who toured the Highlands and Islands and built brochs to order. That in turn must mean that the local landowners were wealthy enough to afford it – and it isn't hard to imagine how rivalry between them would be a spur to taller and taller brochs. Finds at the broch at Gurness on Orkney show that the local aristocrat who lived there was not merely active locally. Fragments of Roman amphorae, or wine carriers, link Gurness to a recorded visit of submission by a Pictish king to the Emperor Claudius at Colchester in AD 43. Whoever the Picts were, they were certainly not primitive relics of the Stone Age.

There has always been a lingering question about what language the Picts spoke. For many years, some linguists believed they may have spoken an ancient tongue unrelated to the Indo-European family which embraces almost all other European languages. To me, and probably to you, while I can see there might be a family connection between, say, Italian and Spanish, it is not at all obvious that German, Portuguese and French are all related. However, be assured that they are and that, along with practically every other European language and others, like Sanskrit from the Indian subcontinent, their grammatical structure shows that they have evolved from a common root. The only living exception in western Europe is *Euskara*, the language of the Basques of north-east Spain and south-west France. Euskara is totally different from any other European language. The grammar is different and the words have quite different roots.

Some scholars, keen to mark out the Picts as an ancient people, relics of the Old Stone Age, have pointed to the few examples of ambiguous runic carvings as evidence that they spoke a language unrelated to any other Indo-European tongue. The truth is that, unless we have more evidence, we may never know. Since there are no written Pictish texts, and so very few surviving stone inscriptions, the language of the Picts enters the realms of the unknowable. Which all adds to the mystery.

Unfortunately, there is virtually no guidance from mythology as to the origin of the Picts. Unlike the rich mythologies of the Irish, the Welsh and the English, the mythology of the Picts is almost non-existent. That does not mean it *never* existed; it surely must have done. It is more a reflection of the absence of writing or, more accurately, the absence of anyone else to write it down until it was too late. In the rest of the Isles, it fell to Christian monks to record, or rather to mould, oral myths. For some reason, this did not happen with the Picts, even though they were among the first people in Britain to have been converted to Christianity after Columba arrived from Ireland in the mid-sixth century. However, the majority of modern scholars now consider that Pictish was closely allied to the strand of Gaelic spoken throughout the rest of Britain and surviving in modern Welsh. If Gaelic was the language of the Picts, it would have been the P-Gaelic of Britain and not the Q-Gaelic of Ireland. If so, then why is the Scottish Gaelic spoken in the Hebrides and now taught in schools in the Highlands and Islands so closely allied to Irish Q-Gaelic and not to the harsher P-Gaelic of the Welsh? For the answer we must look to the west, and to the peninsula of Kintyre, the long finger that reaches almost as far as Ireland itself.

Across the sea from Kintyre, in County Antrim, close to the Giant's Causeway, the Irish kings of Dál Riata began to look for new conquests, and the lands visible across the sea were the natural target. In the first centuries of the first millennium AD, the Dál Riata founded three colonies – on the islands of Islay and Mull and on the Kintyre peninsula. They called their possessions *Ar-gael* – hence Argyll.

The Picts briefly regained Argyll in the sixth century. When Columba arrived we know that it was a Pictish king who gave him the land on Iona in 563. Shortly afterwards, the Dál Riata got a new king, Aidan, who set out to re-establish the colonies in Argyll. If this wasn't enough to upset the Picts, he made matters worse by attacking their possessions on Orkney and on the Isle of Man. He also annoyed the Ui Neill High King of Ireland by these unauthorized adventures. Matters came to a head in 575 when Columba, himself a member of the Ui Neill clan, arbitrated the treaty by which Aidan agreed to pay the High King a military tribute while keeping his maritime revenue for himself. To make the most of this outcome, Aidan built up a strong navy, which is just as well, because he lost most of his land battles. The treaty of 575 kept the peace in Ireland for fifty years, but the Dál Riata never fully recovered their Irish possessions. Their centre of power switched to Argyll and their territorial ambitions were directed north and east towards the lands of the Picts.

For the next two centuries the balance of advantage see-sawed between the Gaels of Dalriada (just another spelling of Dál Riata) and the Picts, with each side alternately gaining ground only to lose it again. Eventually the Gaels gained the upper hand and in 843 the Gaelic king Kenneth MacAlpin was crowned the first king of Alba, a unified country covering both the land of the Picts and Dalriada. Kenneth MacAlpin's claim to the throne was a combination of the Gaelic patrilineal succession of Dalriada and the matrilineal inheritance system of the Picts. These rules did not mean that women became rulers themselves, but that a man would be able to claim the throne through his mother's genealogy rather than his father's. The land became called Scotland because *Scotti* was the label the Romans gave to all Irish immigrants into Britannia. As we have already seen, that name has its own, deeper origins in the mythology of Scota, wife of Mil.

The unification of Scotland under a single king came shortly after the Vikings began their attacks on the coast and is widely seen as a response to this external threat, when unity against a common enemy was more prudent than being weakened by continued feuding, a solution that eluded the Irish. Kenneth MacAlpin moved his centre of operations from Dalriada to the Pictish capital near Perth on the eastern side of Scotland. To emphasize that he was there to stay, he brought the ancient 'Stone of Destiny' from the west and installed it at Scone, near Perth, for his coronation. These decisions, no doubt diplomatically and politically sound at the time, did mean that the centre of power shifted away from the Gaelic west. In later centuries Argyll and the Hebrides consistently refused to be governed by the kings of Scotland, and even now still see themselves as different.

Kenneth was the first of a dynasty of Scottish kings that ruled in patrilineal succession until 1286. Towards the end, Robert the Bruce emerged victorious from a confusion of claimants. His grandson Robert II, the son of Walter, the High Steward of Scotland, and Bruce's daughter, began the Stuart dynasty, which ruled in Scotland until 1603. This was when James VI, on the death of the childless Elizabeth I, also became King of England, and, though it is often forgotten, King of Ireland as well.

The Stuarts were not Scottish in origin at all, but Anglo-Normans. Just as territorial ambition had spurred Richard de Clare, Earl of Pembroke, to invade Ireland in 1166, so other Anglo-Norman lords had their eyes on Scotland. However, unlike Ireland, where the chaos of rival kings made it easy for de Clare to divide and conquer, the relative stability of the unified Scottish royal house required more subtle tactics. Anglo-Norman barons sided with the Scottish kings against the unruly Gaels of the west and it was the contingent of armoured Norman knights on horseback that defeated the Celtic chieftain Somerled's attempted invasion of Scotland at the battle of Renfrew in 1164. Walter the Steward, whose son was to become, as Robert II, the first of the Stuart dynasty, was himself a member of the Norman Dapifer family from near Oswestry in Shropshire, where they had been granted land by Henry I. This is relevant in the genetic context because, although there was no invasion as there had been in Ireland, the Anglo-Norman presence in Scotland was very influential. It may have affected the nature of the Y-chromosome pattern that we find in much the same way that Gaelic and Anglo-Norman Y-chromosomes are distinctly different in Ireland.

So far, we have four possible influences on the genetic structure of the people of Scotland: firstly the Picts; then the Gaels of Ireland, synonymous with the Celts; the Vikings; and, in the south of Scotland particularly, the Anglo-Normans. As we shall see later, the south of Scotland was originally the British Celtic kingdom of Strathclyde. It is known that from the fifth century AD onwards this came under pressure from Anglo-Saxons, but we will leave that to a later chapter. With the Picts, Celts, Vikings and Anglo-Normans to sort out, there is already more than enough to keep us occupied.

12

THE DNA OF SCOTLAND

'We have just had a message,' the pilot's voice came over the intercom, 'that Sumburgh is fogbound.'

It looked as though our trip to the Shetland Isles was going to be cancelled. 'But we'll carry on and see if we can find a gap in the clouds.'

This was not the sort of thing that I, not at my very best in the air, really wanted to hear. We were heading north from Edinburgh airport over the thick layer of sea mist that was covering Scotland for as far as the eye could see. The plane was not a regular jet, but a twin-engine propeller plane, small, cramped and very noisy. Strangely, though, because the plane was small and was driven by propellers, it felt as though all of us, passengers and crew, were part of an adventure. Sure enough, when we reached the Shetlands, after a couple of circuits, the pilot did find a gap in the clouds and he dived through it to make a perfect landing. The team for Shetland was made up of Jayne Nicholson and Sara Goodacre from my research group, my son Richard, then aged eight, whose half-term it was, and me.

For the week we were in Shetland it never really got dark at all. The sun rose at 3.30 in the morning and set at 10.30 at night. But even for the five hours that the sun dipped below the horizon, everything was illuminated by the ethereal northern twilight. It is easily light enough to walk around, and even to read a newspaper, right through the night. And everywhere the air was full of the sound of birds, the raucous clatter of terns and kittiwakes on the coasts and the sweet bubbling of curlews across the moorland away from the sea.

Shetland is nowhere near as fertile as Orkney, but both places – to an outsider – are very different from anywhere else in the Isles. There is a tangible air of Scandinavia about both archipelagos, stronger in Shetland than in Orkney, but unmistakable in both. And it isn't just obvious things like the 'Viking Coach Station' in Lerwick, the capital of Shetland, or signs in shop windows saying 'Norwegian spoken'. It is there in the domestic architecture – the wooden A-frame houses painted with the same rust-red shade that is everywhere in Scandinavia. It is in the undemonstrative, no-nonsense feeling of the place. Although, even recently, anthropologists have written that in Lerwick, and in Kirkwall, the capital of Orkney, practically everyone they saw was blond, I have to say that was not my experience of either place. I could not see an overwhelming presence of the blond Scandinavian archetype which is such a feature of John Beddoe's descriptions, but then I haven't seen it in my visits to Norway and Sweden either. However, the reason we had come to Shetland was not primarily to gaze at the exterior features of the islands' inhabitants but to look for evidence of history hidden from view, hidden in the DNA.

Unlike the rest of Scotland, where most of our DNA samples came from blood-transfusion donor sessions, there are none of these in Orkney or Shetland so we had to arrange other methods of getting our samples. Jayne Nicholson discovered that the Shetland Science Festival was being held in May, so she arranged for us to have a stand at the Festival and also organized a series of visits to schools for the same week. This worked extremely well: while two people manned the booth at the Festival, the others went to schools around the islands. The Festival itself was held in a smart new sports hall on the outskirts of Lerwick, one of many around the islands. The same is true of schools, all of which have brand-new buildings. Shetland Council spends a lot of the revenue it gets from the Sullom Voe oil terminal on upgrading the island infrastructure. The roads are excellent, the inter-island ferries are well equipped and run on schedule. I saw neither poverty nor extravagant wealth on Shetland.

The Science Festival was a jolly affair. Groups from all over Scotland, including a strong contingent from the University of Aberdeen, put on displays of such varied nature as an artificial tornado generator, a giant bubble machine and a practical course in making plaster casts of fossils. Although the Festival was aimed primarily at schoolchildren, there was a healthy flow of adults coming to our stand and we had no difficulty enrolling volunteers in the Genetic Atlas Project. We were not taking blood, only cheek swabs, and I am sure that helped. For this we use a small brush like a miniature bottle-brush, 1 inch long at the end of a 5-inch plastic handle. The bristles on the brush collect cells from the inner cheek as they are rubbed gently over the surface. There is plenty of DNA in these cells and the brushes can be stored for weeks, or posted, without the DNA suffering. It is one of those seemingly unimportant practical changes that actually make all the difference. Now, instead of collecting blood samples, we can send brushes to anywhere in the world and receive DNA back through the post. These brushes can hold DNA safely under even the most extreme conditions. I have equipped a number of university expeditions with DNA brushes and nearly all of them are returned with the DNA intact, even when they have been carried for weeks in a rucksack through deserts and across mountains.

When we began to use the brushes to collect DNA, we would often help the volunteers by doing it for them. However, we have now had to stop this, and volunteers must do it themselves. This is not for fear of breaking new Health and Safety regulations but for a far more delicate reason. False teeth. It was at the Shetland Science Festival that I learned this painful lesson. An elderly lady, eager to join in the project, opened her mouth to allow me to rub the inside of her cheek with the brush. No sooner had I begun to guide the brush across her cheek than it suddenly stopped moving. I let go. I looked at her. She blushed and turned away. After regaining her composure, she returned with the brush and the explanation. I had inadvertently dislodged the top set of her dentures, which had dropped down and clamped the handle of the brush to the lower set. After this, we let people do their own brushing.

The Science Festival was also the scene of another humiliation. One of my obligations in exchange for the display space was to give a public lecture during the Festival, to which I was happy to agree. As usual I spent the previous evening preparing my talk and organizing my slides. With five minutes to go before my talk, I went over to the screened-off section of the hall that had been set aside for public lectures. There was no one there. I checked the time on the Festival programme. This was definitely the right time, and the right place. I waited, but still nobody came, so I thought there must have been some sort of rescheduling that I had not got to hear about. As I was unloading my slides, a lady came in and sat down. I asked if she had come to hear my lecture. She had. Having no audience is bad enough. Having an audience of one is far worse. Unable to slink quietly away, I was honour-bound to give the lecture, all forty-five minutes of it, slides and all. The sole member of the audience sat there quietly, paying attention and, when I had finished, she picked up her handbag and, without a word, left the area. Field work is full of surprises.

One last reflection of Shetland came from talking to men and women at the Festival. I wanted to know whether they felt closer to Scandinavia or to Scotland. On this question, the answer was clear-cut. It was Scandinavia without a doubt. Very few felt any connection with Scotland, let alone with the new Scottish parliament in Edinburgh. It was as if they even preferred to have their affairs governed from Westminster than from Edinburgh. This allergy to Scotland extends to their own individual desires for a Viking ancestry, especially among the men. Where there was any uncertainty, and most people did not know where their ancestors had come from, they wanted to be Vikings. Scots came a very poor second and, to my surprise, an Irish ancestry was even worse.

I know hardly anybody from among my friends and colleagues who has been to Shetland. Only one, who is on the maintenance staff at the Institute where I work, visits regularly. He goes there to witness the festival of Up Helly Aa. This annual event is held on the last Tuesday in January, in the depths of winter darkness, and is a very real reminder of Shetland's Viking affiliations. The day begins with the year's elected Jarl, or leader, and his fifty-seven-strong retinue of *guizers* marching through the streets of

Lerwick dressed in scarlet velvet, wearing winged helmets and carrying elaborate shields and heavy war axes. Becoming the Jarl of Up Helly Aa is a great honour for a Shetlander. It is the culmination of an induction and selection process that can last twenty years and that begins as a teenager with a minor role in the pageant. The Jarl assumes the name Sigurd Hlodvisson for the day and receives the freedom of Lerwick for the duration of his twenty-four-hour reign. The culmination of Up Helly Aa is the ceremonial torching of Sigurd's galley *Asmundervag*, specially made for the occasion. The real Sigurd Hlodvisson, also known as Sigurd the Stout, lived from 980 to 1014. Sigurd was the Norse Earl of Orkney and divided his time between visiting his overseas dominions, in Ireland and the Isle of Man, and summers spent raiding the Hebrides and the Scottish mainland. His reign as Earl of Orkney came to a sudden end when he was killed at the battle of Clontarf when, as you may recall from the last chapter, Brian Boru finally forced the Vikings out of Ireland. Maybe that is why the last thing a Shetland man wants to be thought of is Irish.

We left Shetland after a busy week. The take-off was as alarming as the landing. In certain wind conditions the elected runway faces south, straight at the cliff of Sumburgh Head. I thought we were taxiing to another part of the airport, when the engines went to full throttle and we accelerated down the runway – straight towards the cliff! Needless to say we banked sharply as soon as we left the ground. We had with us over 600 DNA samples, a magnificent total.

We realized that we had the best chance yet of finding genetic evidence of Viking settlement in the Northern Isles. If we could detect the signal anywhere it would be in Orkney and Shetland. And if we could only identify Viking DNA in the Northern Isles, we could look for it elsewhere. According to everything we are told about the Vikings, and which is reinforced by re-enactments like Up Helly Aa, this was a society dominated by warlike chieftains and their blood-thirsty acolytes, raping and pillaging their way around Europe. These being men, it seemed only sensible to begin our search for Viking genes with the Y-chromosome. If these stories were true, then that would be the place to look. Recalling the Irish results, where almost 100 per cent of men with Gaelic surnames are in the same Y-chromosome clan – that of Oisin – what is the paternal clan make-up of Orkney and Shetland? The answer is, very different from Ireland. Although there are still plenty of Oisins in the Northern Isles, the proportion is very much lower than it is in Ireland. Even so, Oisin is still the major clan in both Shetland and Orkney, with just under 60 per cent of men in this paternal clan. That is a very big difference from the situation in Ireland, so this was a very promising start. Almost all the remaining 40 per cent was made up equally of the two clans Wodan and Sigurd, with just a smattering from the minor clans Eshu and Re.

Even without delving any further into the detail of the Y-chromosome genetic fingerprints, it was clear that Ireland and the Northern Isles had a very different genetic history if we listened only to the version told by men. But it wasn't completely different. Oisin still dominated, as it did in Ireland, but nowhere near as much. Our first thought, when we saw these results, was to draw the conclusion that in the Northern Isles Oisin represented the descendants of the indigenous Pictish ancestry, while the men in the clans of Wodan and Sigurd had Viking ancestors who had come from Norway. That would put the ancestral proportions in present-day Shetland at roughly 65 per cent Pict and 35 per cent Viking. The indigenous Pictish ancestry would still be in the majority but with a big slice of Viking male ancestry.

The first test of this theory was to see what things were like in Norway. To prepare for this comparison, two of the team, Jayne Nicholson and Eileen Hickey, had already been collecting in Norway. Thanks to the co-operation of the Norwegian Blood Transfusion Service, we had 400 blood samples from all over the country, from Finnmark in the far north to Rogaland in the extreme south. If Norwegian Y-chromosomes were all either in the clans of Wodan and Sigurd, with no Oisin, then it would back up this first conclusion – at least to the coarse level of detail embraced by simple clan membership. However, as it turned out, there were plenty of Oisins in Norway as well. Altogether, nearly a third of Norwegian men were members of the clan of Oisin. The straightforward link between Oisin = Pict and Wodan and Sigurd = Viking that we had begun to hope for in our first run through the Shetland Y-chromosomes had obviously been an oversimplification.

However, when we looked at the clan make-up in the different regions of Norway, the concentration of Oisins in the western provinces, the traditional homeland of the Vikings, was much lower than in other parts of Norway. Around Bergen, on the south-west coast, only 15 per cent of men had Oisin clan Y-chromosomes. The other two thirds of Norwegians were split between the clans of Wodan and Sigurd, with Wodans out-numbering Sigurds by roughly two to one.

We were faced with two questions before we could be sure of interpreting our Shetland results correctly. The first was this. Were the Norwegian Wodans and Sigurds genetically similar, at the more detailed fingerprint level, to the men in the same clans from the Northern Isles? In other words, did we find the same detailed Y-chromosome fingerprints in Norway and the

Northern Isles within each of the clans? We checked each one, looking for matches in the Norwegian men. To our great relief, we found exact or first-generation matches to almost all the Sigurds and to about two thirds of the Wodans. This certainly looked like a good indication that most, if not quite all, of the Y-chromosomes in these two clans had arrived from Scandinavia. But how about the Oisins? Here again there were matches between the Norwegian and Shetland samples, but nowhere near as many. There were far more Shetland Oisins whose Y-chromosome signatures were unmatched in Norway than in the other clans. We expected some similarities, since the Viking ships would not have distinguished between genetic clan when choosing their crews and there were plenty of Oisins in Norway. When we took this into account it was clear that our initial estimates of Viking ancestry had been a bit low. Some of the Northern Isle Oisins had almost certainly come with the Vikings. Including them pushed our estimate of Viking ancestry in men from Shetland up to 42 per cent. The proportion of Orkney men with a Norse Viking ancestry, which we estimated in the same way, was slightly lower, at 37 per cent. But please do not concern yourself with exact proportions; just take from this that the male Norse ancestry of Orkney and Shetland is substantial, but was never complete.

I began our interpretation of the Northern Isles DNA with the Y-chromosome because of the Vikings' reputation, but I also wanted to see what the maternal DNA told us as well. At the time we were analysing the DNA from the Northern Isles we had just completed a study with Agnar Helgason, an Icelandic anthropologist, on the genetic ancestry of his native land where we had asked a similar question about the paternal and maternal input. The histories of Iceland and of the Northern Isles were quite different in that, when the Vikings began to settle Iceland from around AD 860, it was uninhabited. The few Irish monks who had settled there in the quest for a life of contemplation sensibly left as soon as they saw the first sails on the horizon.

Over the next fifty years large numbers of Norse settlers arrived in Iceland, most from around Bergen but some from Viking settlements in Britain. This was large-scale, planned immigration to a land with no indigenous opposition and by the beginning of the tenth century there were 60,000 people living in Iceland. The population has grown to 250,000 today, but there has been no recorded large-scale immigration since the original settlement. Agnar and I wanted to find out whether this had been a predominantly male-driven settlement, with females brought in from elsewhere, or whether roughly equal numbers of Norse men and women had arrived. There have been persistent stories that Icelandic men raided the coasts of Scotland and Ireland for wives. Agnar and I thought we could check these stories by comparing modern Icelandic Y-chromosomes and mitochondrial DNA with the equivalent DNA from Scandinavia, Ireland and Scotland. By going through the Icelanders' DNA results one by one, we assigned their most likely origin from comparison with British, Irish and Scandinavian samples.

We discovered that roughly two thirds of Icelandic Y-chromosomes were Scandinavian, while the remaining third were from Ireland and Scotland. However, the origin of maternal DNA was reversed, with only a third from Norway and two thirds from Ireland and Scotland. This confirmed the stories that, while most of the men had settled in Iceland from Norway, they relied heavily on women imported from Ireland and Scotland. It doesn't necessarily mean they were taken there against their will, as the results could not distinguish between settlers who had arrived straight from Norway and the male descendants of Vikings who had spent a generation or two in Scotland. Even so, it is hard to account for the Gaelic origins of a third of Icelandic Y-chromosomes without contemplating that these men were taken to Iceland as slaves. The Iceland study gave us a very interesting result, and also helped us to develop the way of assigning the Icelandic DNA to a Norse or Gaelic/Pictish origin, which we then used for Orkney and Shetland and then, in modified form, for the rest of the Isles.

When we tried the same treatment on the Northern Isles, expecting a similar result to what we had found in Iceland, we were in for a major surprise. The maternal clans in Norway and in Orkney and Shetland were superficially quite different. Katrine was common in Norway but rare in Orkney and Shetland, and the same was true for Tara. But when we looked more carefully at the detailed sequences, the matches leapt out. Within each of the seven major clans, and the minor ones too, the similarities in detailed sequence were remarkable. I had initially expected to find hardly any Scandinavian mitochondrial DNA in either Orkney or Shetland. I had imagined that the Viking reputation for rape, pillage and general destruction recalled in Up Helly Aa, in atmosphere if not in fact, would have had the expected genetic consequence – kill the men and keep the women. When it came to permanent settlement, these same women, I had expected, would have become the wives of Viking men. That is the usual pattern of conquest and settlement that I have seen many times throughout the world. It is all too obvious in the genetic consequences of the European colonization of Polynesia and South America, where European Y-chromosomes are extremely common, while European mDNA is virtually unknown. This is a record of great success for the incoming Y-chromosomes at the expense of the indigenous, but with no effect at all on the aboriginal mitochondrial DNA. Orkney and Shetland had all the right ingredients, but the genetics said otherwise. Amazingly, there was as much Norse mitochondrial DNA in the Northern Isles as there were Norse Y-chromosomes. This could mean only one thing – anathema to the Jarls for the day of Up Helly Aa and their retinue of axe-wielding *guizers*. The Viking settlement of Orkney and Shetland had been peaceful! The Scandinavians had brought their women with them.

The 60 per cent of Orcadians and Shetlanders who do not have a Viking genetic ancestry are most likely to be the descendants of the indigenous Picts. However, there is a proviso. After the islands were eventually ceded to Scotland in the fifteenth century there was a substantial immigration of Scots, which would have diluted the genes of the islanders, whether of Viking or Pictish ancestry. Since we had been successful in identifying Viking genes, both male and female, the next question was whether we could do the same for the Picts, and for that we must head south to the heart of Pictland.

Close by the small town of Dunkeld, a few miles north of Perth on the banks of the River Tay, is the site of the Abbey of Scone. It was here that Kenneth MacAlpin was crowned as the first king of a united Scotland in 843. The area around Dunkeld was the central stronghold of the Pictish kings and Kenneth, a Gael from the west, deliberately chose Scone for his coronation to symbolize the unity between Pict and Celt which his reign proclaimed.

Beneath the coronation throne lay *Lia Fail*, the Stone of Destiny, a rectangular block of sandstone. It is said that *Lia Fail* could talk and that it spoke the name of the next king. The Stone itself has a mythical history linking it to Egypt, Spain and Ireland, reminiscent of the Irish origin myths of Mil. However, geologists who have examined the Stone say it comes from the neighbourhood of Scone itself. But there is an explanation for that. The kings of Scotland continued to be crowned above the Stone of Destiny until 1296 when Edward I, always aware of the power of symbolism, carried off the Stone and installed it in Westminster Abbey. But, according to the legend, he was duped. Monks from the abbey, warned of the approach of Edward and his army, hid *Lia Fail* nearby and replaced it with a slab of local sandstone. It was this replica which Edward took back to England while the real Stone lies hidden somewhere close by.

This neatly explains the geological similarity of the Stone to local rocks, and also why there continued to be a long succession of Scottish kings even when the Stone was lying in England. It could hardly be expected to speak the name of the next king of Scotland if it was installed in Westminster Abbey. Why didn't the monks recover the Stone from its hiding place once Edward had departed? For fear of retribution once he knew he had been tricked is the rationalization of the myth. By the time it was safe to bring it out of hiding, the monks had forgotten where they put it. In the eighteenth century there was a local legend that, after a violent storm, a farm lad discovered an underground cavern which had been exposed by a landslide triggered by the torrential rain. The lad entered the cavern and found a stone covered with inscriptions, as indeed the original *Lia Fail* was recorded to have been. Thinking it of no importance he did not speak of it until years later when he heard the story that the monks had switched the stone. Alas, when he returned to the spot, he could not find the cavern entrance, presuming it to have been once more covered by a landslide. This all sounds very unlikely, but stranger things have happened and I am reminded how the prehistoric caves at Lascaux, in the Dordogne, the walls of which are covered with the ancient paintings of bison and reindeer that our ancestors hunted 20,000 years ago, were discovered by accident by another farm lad at about the same time. So perhaps *Lia Fail* really is still there, waiting to be rediscovered.

Meanwhile, the Stone in Westminster Abbey remained resolutely where it was, beneath the coronation throne for every English monarch since Edward II, up to and including the present Queen Elizabeth II in 1953. That will be the last time, for in 1996, 700 years after it was taken to London, *Lia Fail* was returned to Scotland. In an elaborate procession along the Royal Mile, lined by 10,000 people on St Andrew's Day, the Stone was taken from the Palace of Holyroodhouse to its new home in Edinburgh Castle. To the sound of a twenty-one-gun salute from the castle ramparts, the Stone was laid to rest in the Great Hall. The strength of feeling which energized the campaign to return *Lia Fail* to Scotland after 700 years was formidable. The ceremony which attended its return was in many respects the assertion of an ancient Pictish connection, in the same way that Up Helly Aa celebrates the Norse identity in Shetland.

From a genetic point of view, I wanted to see whether I could find a parallel in the living descendants of the ancient Picts. Was there, hidden deep within the cells of Scots still living in the Pictish heartland, a signal of their ancient identity every bit as real, or perhaps more so, as the Stone of Destiny itself? We began our search for Pictish genes at Auchterarder, 15 miles south-west of Perth and temptingly close to the famous golfing hotel of Gleneagles, but I am sad to report that our research budget did not stretch to that level of subsistence. Auchterarder was the first of many visits that my research team paid to blood-donor sessions.

Three months before, in the spring of 1996, I had spent a week travelling all round Scotland visiting the directors of all the Scottish Blood Transfusion Service centres, enlisting their help in our project. It was never difficult to explain why we wanted to do this work, but there were a lot of details to be sorted out in getting permission from the donors' representatives, as well as formal permission from the Transfusion Service itself, ensuring we did not compromise the confidentiality of the donors. We also had to agree a way of collecting the blood that would not interfere with the smooth running of the donor sessions. There was one thing both I and the directors were agreed on. We must attend the sessions in person. Too many researchers ask for blood to be collected on their behalf, without actually going to the sessions. This makes extra work for the donor nurses. I also wanted to be sure we were there to explain our project to the donors and get their consent, and also to talk to them about their own backgrounds and to get the feel of the place.

You may be a blood donor yourself, in which case you will know how the sessions work. As each donor arrives, they wait to be checked in. This, we jointly decided, was the best time for us to introduce ourselves and to ask for their consent to having us analyse their DNA. I was extremely fortunate in having a team of researchers in my group at the time who were absolutely brilliant in the art of persuasion. To the waiting donors, we explained that we were creating a new genetic map of Britain and trying to work out from it our Celtic, Pictish and Viking roots. That was about all the explanation anyone needed before agreeing to take part, especially when they realized we would only be taking a sample of blood from their donation, so there was no need for another needle. Donors told us where they were from, as far back as their grandparents. Those who didn't know, generally the men, took a purple form home with them to ask someone in the family, invariably a woman, and sent it on to us.

By the end of the session in Auchterarder we had collected 187 blood samples, a wonderful start. Over the next two years we visited almost all of the donor sessions throughout Scotland, from Galashiels and Thornhill in the Borders, to Thurso at the very top of Caithness, to Stornoway in the Western Isles, to Campbeltown on the tip of the Kintyre peninsula. Everyone in my research lab joined in, even if they were working on different projects. Itineraries were prepared so that, with luck, several different sessions in different regions could be covered in one trip. We travelled to Scotland by road, by air or by the Highland sleeper from Euston. We set aside a small office at the Institute as a planning room, with donor-session schedules on the wall and a large map of Scotland next to them. By the time we went to our last Scottish session, two years later in Fort William, we had collected over 5,000 blood samples and clocked up over 50,000 miles between us. It is a testament to the team's powers of persuasion that we only ever had one person decline to take part in the project – a farmer from Callander in the Trossachs, north of Stirling – who had to rush off because one of his cows was about to give birth and he didn't have time to fill in our form. It says a lot about him, and donors in general, that they take their blood donations very seriously. None of the donors is paid a penny and the sessions have a tangible atmosphere of selfless community service. Most sessions are entirely run by women, with the only men present being the drivers of the vans that bring the teams and the equipment. It was friendly, calm and efficient. Very impressive all round.

To begin looking for the genetic signatures of the Picts on the mainland I began by dividing Scotland into regions. It was easy to decide where to draw the line on Orkney, Shetland and the

Hebrides – they were islands – but on the mainland I needed to draw boundaries. These are shown on the map on page xiii. Pictland was covered by two of these: the Grampian region and Tayside and Fife, which for convenience we will henceforth refer to simply as Tayside.

Since we began our analysis of the Northern Isles with the Y-chromosome, and also in deference to the Pictish tradition of matrilineal inheritance, we began the search with mitochondrial DNA. The clan proportions for both Pictish regions were remarkably similar to one another – again you can see this in the Appendix. When I put the results through a statistical test, the only clans that had a significantly different frequency in the two regions were Jasmine, higher in Grampian, lower in Tayside, and Tara, higher in Tayside, lower in Grampian. Otherwise there was no difference. It looked, at this level of scrutiny, as if the maternal ancestry of the two Pictish regions was almost indistinguishable one from the other.

When it came to analysing the detailed sequences, though, I could see plenty of differences. That is always to be expected, because a large proportion of mitochondrial sequences are unique to one region. At the risk of being technical, of the 170 different sequences we found in the two Pictish regions of Grampian and Tayside, 70 occurred only once. I used our experience with disentangling the Norse and Pictish components in the Northern Isles to devise a simple score between 0 and 100 which would summarize the similarity between the two regions. If all the sequences in one region have an exact match in the other, then the score is 100. If none of them matches, the score is 0. On this scale the match score for Grampian and Tayside was 77. Of course that doesn't mean a great deal without anything else to compare it with but, as we shall see, 77 is a very high score compared to most others. It is high enough to consider both regions as one for the purpose of searching for Pictish genes.

When I began to look in detail, it was immediately obvious that the DNA from these two regions was not really all that different from much else in Britain. There was no sign at all of the exotic sequences one might associate with a truly relic population that had been somehow isolated from the rest of mankind. Lots of sequences were unique to the two regions, but that isn't unusual, as we have just seen. Looking through these unique sequences, I could see they were closely related to mDNA sequences from other regions, differing from them by just a single mutation. There was nothing very special about the Pictish DNA, at least on the maternal side. It didn't seem to me that, on this evidence, a case could be made for treating the Picts of Tayside and Grampian as being particularly unusual. But that was just how they appeared to me at the time. We would certainly need to compare them with the rest of the Isles to gauge their true nature. That was my impression from the maternal signal, but what would the Y-chromosomes look like?

Once again the overall Y-chromosome clan structures in Grampian and Tayside were, like the maternal signals, remarkably similar to each other. The clan of Oisin predominated in both, rising to 84 per cent in Grampian – not quite as high as the west of Ireland, but much higher than in Orkney or Shetland. Wodan was quite high in both, at 12 per cent in Grampian and 18 per cent in Tayside respectively, but Sigurd was very low indeed. Only 2 per cent of men in both Pictish regions belonged to Sigurd's clan. You will recall that, in Orkney and Shetland, we assigned all the Sigurds to a Norse Viking origin. On the evidence from the Pictish regions, with low numbers in the clan of Sigurd, it looked as if Grampian and Tayside had virtually no Viking ancestry. This is precisely what we would have expected from the history and the archaeology of both regions. There are no remains of Viking longhouses and no Norse place-names. In fact, some of the place-names have recognizably Pictish origins, notably Pitlochry on the River Tummel a few miles north of Dunkeld. In Orkney and Shetland the reverse is the case. All the place-names have Norse origins.

In Pictland, the genetics suggests a very low level of Viking ancestry among the men. However, if we accept that, as I think we should, what can explain the substantial percentages of Wodan in both regions? In the Northern Isles the proportions of Sigurd and Wodan were roughly equal. If, as we had done, we attribute both clans in the Northern Isles to Viking settlement, based on the close affinities with Y-chromosomes we know exist in Norway, how do we explain the Wodan presence in Pictland? If Viking settlers in Orkney and Shetland were composed of roughly equal numbers of Wodans and Sigurds and these reflected the composition of a typical Viking immigration any-where in the Isles, then only 2 per cent of the Pictland Wodans could have a Viking origin, leaving the other 12–16 per cent un-accounted for. When I checked through the detailed signatures of Pictland and Northern Isles Y-chromosomes from the clan of Wodan, there were plenty that matched – and plenty that didn't. This was a puzzle. The Pictland Wodans could not all have arrived as Vikings, but where had they come from? Certainly not Gaelic Ireland, where they are almost unknown. Perhaps these men in the clan of Wodan really were the surviving descendants of the Picts.

I made the settlement date calculations, as I had in Ireland, for both the paternal and the maternal ancestors. There was still a wide gap between the male and the female side. The ages of the mDNA clans varied between Ursula at 9,200 years, slightly older than in Ireland, and Jasmine, again the youngest at 5,000 years. The paternal clans were slightly older than in Ireland, but still much younger than the maternal dates. There was no immediate answer to what this meant; indeed by this time I had more or less decided to wait until I had surveyed the entire Isles before trying to make sense of it all. But to what extent had the Picts been replaced by the Dalriadan Celts, the Gaels from Ireland? To try to find out we need to move away over the mountains that separate Pictland from our next destination – the Celtic west.

The gradual colonization of the west from the Irish kingdom of Dál Riata during the first half of the first millennium AD, and the consolidation of their Gaelic kingdom in Scotland following their defeat by the Ui Neill, had an immense cultural impact in Scotland. As we have seen, the language changed from the P-Gaelic of the Picts to the Q-Gaelic of the Irish and, on the accession of Kenneth MacAlpin in 843, Gaelic political dominance was complete.

In the west, one question which our genetic analysis could hope to answer was the degree to which the arrival of the Dalriadan Celts from Ireland displaced the Pictish inhabitants of the region. But there is also another factor to bear in mind, and that is our old friends the Vikings. The whole of the west coast and the Hebrides were repeatedly raided by Vikings from the first attack on the monastery of Iona in 795. If the experience in Orkney and Shetland is anything to go by, we should expect to see evidence of a Viking presence among today's inhabitants of the west coast and the islands.

When we were collecting our samples from the west coast and from the Hebrides, there was a distinctly different response to my questions about people's own ancestry. In Shetland the last thing men, in particular, wanted to see in their genes was the signal of an Irish ancestry. In the Western Isles, and along the west coast, there was still a certain arcane thrill at the possibility of a Viking ancestry, but this was eclipsed by an affiliation with a Celtic past, whatever that actually meant. People were keen to expose an Irish ancestry, if there was one, but most showed no real interest in the prospect of being of Pictish descent. And yet that was the most likely outcome, since it is almost always, in my experience, the earliest occupants who dominate the gene pool of a region. The later arrivals may get all the headlines, but it takes a lot to displace indigenous genes, especially on the female side. Thanks to the Scottish Blood Transfusion Service, we travelled to donor sessions from Thurso in the far north, along the west coast to Ullapool, Gairloch and Fort William, then south to Oban, Lochgilphead at the top of the Kintyre peninsula and right down to Campbeltown at the very end. We travelled over the sea to the Western Isles and across the bridge to Skye. We saw the land in all its moods, from brilliant sunny days when the bright hills shone in the sunlight to furious tempests when wind and rain lashed along streets and through doorways.

Not even the foulest weather prevented the calm progression of the donor sessions, even when the rain was coming in through the windows of community centres that had seen better days. Attendance by the donors was just as high in the bad weather as in the good. In Grampian, almost everyone coming to the donor sessions had been born nearby, and so had their grandparents. In this respect, the most stable place we visited was Huntly in rural Aberdeenshire, where 78 per cent of donors had all four grandparents born close by. In the west this figure was quite a bit lower, and there was a noticeable proportion who had moved into the area in the recent past, mainly from the towns of the central lowlands or from England. Since the project covered the whole of Britain, practically everybody could contribute to the outcome, even if they had only recently arrived in their current locality. It also worked both ways. In and around Glasgow, Edinburgh and London we often encountered donors whose ancestors had come from the west of Scotland and where, for the purposes of the genetic map, they could be confidently placed.

What of the results? We were becoming very adept at identifying Viking DNA and, sure enough, we found plenty of it. In Caithness and along the stunningly beautiful north coast from the Kyle of Tongue to Loch Eribol and Durness we found, by the same tests we had used in Orkney and Shetland, that 15 per cent of the DNA was Norse in origin. Like the Northern Isles, this was true both of Y-chromosomes and of mitochondrial DNA, so it looked as though it was by establishing family-based communities that the Vikings came to settle here, however unlikely this sounds in relation to their folk memory as bloodthirsty plunderers. However, in the Western Isles and Skye, the genetic evidence for a more typecast male-dominated Viking colonization began to emerge when we looked at the results.

There are twice as many Norse Y-chromosomes in Skye and the Western Isles as there are Norse mitochondria; 22 per cent of Hebridean Y-chromosomes, but only 11 per cent of mitochondrial DNAs, had a Norse origin. The further down the west coast, the lower the Viking component became until, in Argyll, it was down to 7 per cent for Y-chromosomes and only 2 per cent for mitochondria.

The diminishing Viking input and its increasing asymmetry between the sexes as we travel down the west coast seems to me best explained by a gradual process of Viking settlement from the early bases in Orkney and Shetland. Some men took their women with them, or returned to Orkney to bring their families once they had laid claim to a plot of land; others intermarried with local Gaelic or Pictish women. In general we found the same detailed Norse Y-chromosomes along the west coast and in the Hebrides as we had already discovered in the Northern Isles. It really didn't look as though there had been a rush of fresh arrivals from Scandinavia.

If that was the level of Norse DNA, what of the rest? Could we assign this to Celtic or Pictish origin? For this, I made a start by comparing our results from the Pictish regions of Grampian and Tayside with the west coast locations. I could tell straightaway that they were substantially different. Not only that – there was also a big difference between the three regions of the west. Much as I had divided Pictland into Grampian and Tayside, so I split the west into three. They were the Highlands from Durness to Fort William, then from Oban south to Kintyre, which I grouped together as Argyll, and, thirdly, the Hebrides, which combined Skye and the Western Isles.

On clan comparisons alone, the Hebrides stood out as very different from the other two. Argyll, at this crude level, was far more like the Pictland regions of Grampian and Tayside than the Hebrides. The Highland coast was somewhere in between. Even when I removed the Norse DNA, the picture was the same. At the greater level of detail revealed by the precise sequences, in the case of the mitochondria or the profiles of the Y-chromosomes, the stark differences between the regions still stood out. For the mitochondrial comparisons, on the scoring system of similarity that I introduced in Pictland, which goes from 0–100 (the higher the score, the greater the similarity), Argyll vs Pictland scored 60, the same score as Argyll vs Highlands but much higher than the Hebrides scored in this equation of similarity with either of them. However, the Y-chromosomes told a different story. The Argyll Y-chromosomes were much more like their Hebridean counterparts than those in the Highlands.

If your head is spinning, you are feeling just as I did when I first tried to decipher these results. It seemed to be going so well. We had identified the genetic legacy of the Vikings and we had found that, just as the archaeology and history leads us to expect, they did not settle in Pictland to any extent. We had seen their diminishing genetic impact as we travelled further and further away from their forward bases in Orkney and Shetland. Until then everything made good sense. But then, the simple story, based on our historical assumption, began to unravel. Far from Pictland being genetically distinct from the Celtic heartlands of Dalriada and Argyll, they were remarkably close, on the maternal side at least. However, this similarity is not reproduced by the Y-chromosome, where Argyll has a low gene-sharing score with Grampian, even after the Norse component has been subtracted.

To me this is the familiar signal of maternal continuity. What we have here, I think, is the imprint of Scotland's Pictish ancestry, on the maternal side, spread more or less uniformly across the land. This is the bedrock of Scottish maternal ancestry on which more recent events have been overlaid. The maternal gene pool is more or less the same in Pictland, in 'Celtic' Argyll and in the Highlands. In Orkney and Shetland, the Pictish bedrock has been overlain by a more substantial and identifiable Norse settlement than anywhere else in Scotland, but it is still there nevertheless.

On the male side, we can see plainly what must be the Pictish bedrock in Grampian and Tayside, but in Argyll it has been sub-stantially overlain by new arrivals. The Argyll Y-chromosomes are in between the Irish and Pictish values and, although these estimates are approximate, a 30–40 per cent replacement of

Pictish by Gaelic Y-chromosomes would account for this. It is much harder to be accurate in this case than it was in judging the Norse contribution to the Northern Isles because of the basic similarity between Irish and Pictish Y-chromosomes which, incidentally, makes it almost impossible to detect any genetic effect of the Ulster plantations. However, the genetic signal, as far as I can judge, points to a substantial and, by the look of it, hostile replacement of Pictish males by the Dalriadan Celts, most of whom relied on Pictish rather than Irish women to propagate their genes. The reason I cannot be more certain is itself very relevant to the myth of the Picts. It is precisely because they are genetically close to the Gaelic Irish that these estimates are so difficult. If they had been a relic people, a genetic isolate, then it would have been easy to distinguish them from Irish Gaels. But on the contrary, it is extremely difficult, from which we can confidently conclude that the Picts and the Celts have the same underlying genetic origins.

Which leaves the Hebrides. Their genetics stand out from the Picts and the Celts and the Norse of Shetland or Orkney. Their DNA-sharing scores are low for all comparisons, and for both maternal and paternal genes. Take away the attributable Norse component and the differences remain. What can be so special about the Hebrides? Let's take a closer look.

The long line of islands of the Western Isles are battered on their western sides by the pounding of the Atlantic. These islands protect the Inner Hebridean islands of Canna, Rum, Eigg and Skye from the worst of the Atlantic swell – though when I was caught on an inter-island ferry in a gale I thought the violently bucking boat could have done with a lot more protection. The islands are the last stronghold of the Gaelic language in Scotland and had suffered from decades of depopulation even before the

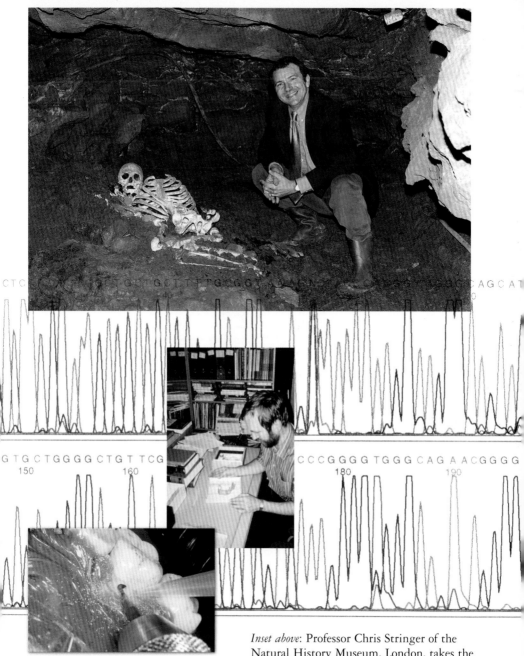

CT C G ... G T G T T G C G G G ... ... A G C A T

G T G C T G G G G C T G T T C G
150   160

C C C G G G G T G G G C A G A A C G G G G
180   190

Top: History teacher Adrian Targett meets Cheddar Man. Tests on mitochondrial DNA showed they are related to each other over a span of 9,000 years.

Middle: The readout from a DNA sequence analyser that provides the raw material for *Blood of the Isles*. Coloured peaks on the trace denote the order of the four DNA bases, A, C, G and T, in a segment of DNA.

Inset above: Professor Chris Stringer of the Natural History Museum, London, takes the 12,000-year-old lower jaw from its storage case prior to my attempts at DNA recovery.

Inset below: Extracting dentine powder, which contains DNA, from a molar belonging to a young man who died over 12,000 years ago in Cheddar Gorge, Somerset. The teeth are in excellent condition and the enamel has protected the DNA during the intervening millennia. The pink material is a dental mastic, which I used to protect the rest of the jaw.

Above: The Priory at Lindisfarne. The ruins occupy the site of the undefended monastery which was attacked on 8 June 793 in the first recorded Viking raid in the Isles.

Below: The beach just below the monastery at Lindisfarne – perfect landing for the Viking longships in advance of their attack.

Left: Viking longship of the type that terrorized the Isles for 500 years. The sleek lines and shallow draught made for both speed and stealth. Longships like this could be swiftly beached on almost any coastline for a lightning raid.

Above: Jarl squad members in full Viking dress celebrate Up Helly Aa in Lerwick, Shetland. The annual pageant, held on the last Tuesday in January, celebrates Shetland's Viking past and culminates in the burning of a replica Norse longship.

Below: Beddoe describes the inhabitants of northern Lewis as a 'large, fair and comely Norse race, said to exist pure in the district of Ness at the north end of the island'. Plainly they are still there.

Right: The 5,500-year-old Neolithic *cromlech* at Carreg Sampson in Pembrokeshire, south-west Wales. Originally a covered grave, the earthen mound has eroded to reveal the megalithic internal structure.

Left: The Roman fort at Caerleon in Gwent. The Romans spent huge amounts of time and effort trying to subdue the Welsh, with only moderate success.

Below: Offa's Dyke, near Knighton in mid Wales, constructed in the late eighth century in yet another attempt to confine the troublesome Welsh and prevent them from raiding England.

Right: The head of Offa, King of Mercia, who considered himself to be in the same league as his eighth-century contemporary Charlemagne.

Left: Pembroke Castle, built by Arnulf de Montgomery, Earl of Shrewsbury, the first of the Norman Marcher Lords to make significant inroads into Wales in 1093.

A young Welsh fan cheers on his team at the 2003 Rugby World Cup finals match against Italy. Merlin's Red Dragon is still a prominent feature of Welsh identity.

The sun streams into the
Neolithic passage grave at
Newgrange, County Meath,
at the midwinter solstice.
The massive tombs clustered
around Newgrange on the
Boyne were constructed
about 5,200 years ago.

Left: The rocky coastline of south-west Ireland, near the river Kenmare, the first landfall for Mil on his way from Spain to defeat the Tuatha Dé Danaan.

Above: The sacred hill at Tara, seat of the High Kings of Ireland.

Left: A delicate model of a sea-going vessel, made of solid gold between 1,900 and 2,100 years ago, and probably a votive offering to Manannan mac Lir, King of the Ocean.

Below: This bronze statue of Cú Chulainn, Celtic super-hero and the saviour of Ulster in the war with Connacht, stands in Dublin's main Post Office, the scene of the most intense fighting during the Easter Rising of 1916.

Below: St Patrick's Day, the most vigorously and widely celebrated of the four national Saint's Days. Here crowds throng Fifth Avenue, New York, for the 2006 parade, the 245th in the city's history.

Above: The resurgence of the Celt. The dramatic entrance to Canolfan Mileniwm Cymru, the newly opened performing arts centre in Cardiff, with two lines of poetry by Gwyneth Lewis, one in Welsh, the other in English.

Below: Once more, young children, like these from Lahinch, County Clare, are having lessons in the ancient languages of the Isles.

Highland Clearances of the eighteenth and nineteenth centuries. The sight of an abandoned village, with the outlines of stone-walled cottages collapsed and overgrown, is always a sad one. But the sheer scale of it only struck me one day last December when I was taking advantage of the few hours of sunlight to have a trip in the car around Skye.

I went, for the first time, to Glen Eynort, near the Talisker whisky distillery at Carbost. The road to Glen Eynort rises from Carbost and then crosses a low pass which leads to the glen with its two or three inhabited cottages, each with a plume of smoke rising straight up into the cold and, for once, still air. As I descended, the sun, very low in the sky, came out and lit up the hillsides which surround the glen. What I saw utterly amazed me. The low angle of the sun transformed every patch of hillside into lines of light and shade. It took me a moment to realize that these were abandoned fields, with the sun striking the ridges and casting a shadow in the furrows. This ancient landscape, glimpsed only because of the low angle of the sun and normally impossible to make out, covered hundreds of acres. The valley must once have been teeming with life but now, save for the three cottages, it was empty, the crofters dispersed to the far-flung corners of the world.

Luckily the depopulation was not complete and the Hebrides are once again thriving. But why are the genetics so unusual? On the maternal side, the striking thing, compared to other regions, is the much higher proportions of clans that, though they certainly occur elsewhere, are much less frequent. These are the two clans of Jasmine and Tara. And these clans carry the unmistakable signature of agriculture. As we saw in an earlier chapter, Jasmine herself lived in the Middle East and her descendants accompanied the spread of farming into Europe. The clan divided into two around the Balkans. One branch followed the Mediterranean and Atlantic coasts, the other crossed Europe overland. The two branches can be told apart by a series of mutations which must have happened after the split. Both branches have the characteristic mutation of all members of Jasmine's clan of 069 and 126, plus two others at 145 and 261. On its way round the Mediterranean, one branch acquired two more changes at 172 and 222, while on its trek through Europe the other branch gained one extra change at position 231. It is still amazing to me that these tiny changes can tell us so much about the journeys of our ancestors. How one line, hugging the Mediterranean coast, reached Spain then headed north up the Atlantic coast of France, while the other forced its way through the forests and valleys of continental Europe. These mutations illuminate such different journeys. In Ireland, and all along the west coast of Scotland, only the Mediterranean branch of Jasmine is found. The same is true of the Hebrides, and the concentration of these Mediterranean Jasmines in Skye and the Western Isles explains one of the differences.

The other difference is in the clan of Tara. When I first discovered the clan, with my colleagues, it had all the characteristics of a hunter-gatherer origin: a founding date of 17,000 years ago, and a location to the north of Italy among the hills of Tuscany.

When my research group were defending our conclusions about the human past in Europe, and had gathered lots more samples in the process, it gradually became clear that the clan of Tara had, within it, a branch that looked much younger than the rest of the clan. As well as the signature Taran changes at 126 and 294, this branch had additions at 296 and 304. We dated this branch at slightly younger than Tara herself, possibly just within the scope of the spread of agriculture, possibly at the end of the

Mesolithic. And it is this Taran branch that dominates the Hebrides.

It seemed to be these two young branches of the two clans, Jasmine and Tara, that were responsible for the unusual genetics of the Hebrides. And both of these branches, the Mediterranean Jasmines and the younger Tarans, had a distinctly seaborne flavour about them. They are spread all along the Atlantic fringe, but rarely inland. This was a definite clue, but the solution was not yet clearly visible. The Hebrides are also unusual in having a very high concentration of members of the Katrine clan, especially on Lewis, where they reach the highest frequency of anywhere in Scotland.

It was during my research on Skye that I stumbled across a genetic phenomenon which, with hindsight, I should have investigated much sooner. At that stage I had already discovered the link between my own surname and a particular Y-chromosome profile. Foolishly imagining that such a link to a common founder would only be found in comparatively uncommon English surnames, it took another year before I realized the same might be true, albeit in diluted form, among Scottish clans as well. My research student Jayne Nicholson and I found a rare Y-chromosome profile in our Skye samples and, when we compared it to others collected at donor sessions else-where in Scotland, we noticed that we found it almost exclusively among men with the surnames Macdonald, McDougall and Macalister. It was Jayne who remarked that all three names were said to be descended, according to traditional clan genealogies, from Somerled, the Celtic hero whom we have already en-countered. It was Somerled who was responsible for ending the power of the Norse earls of Orkney in Argyll and the Hebrides, and who died at Renfrew during his ill-fated invasion of Scotland in 1164. Jayne set about writing to men with these three names asking for DNA samples, while I contacted the five living clan chiefs whose genealogies traced back to Somerled. They all agreed to help and, I'm very glad to report, all five had inherited the same Y-chromosome that we had seen in the men with the three surnames. Somerled's Y-chromosome had done extremely well and, thanks to its association with a powerful and wealthy clan, has become very common indeed in the Highlands and Islands, and among Highlanders who have emigrated overseas. Roughly a quarter of Macdonalds, a third of McDougalls and 40 per cent of Macalisters are direct paternal descendants of Somerled. This is not just true in Scotland, but throughout the world; it has been estimated that there are 200,000 men who carry Somerled's Y-chromosome as proof of their descent from the man who drove the Norse from the Isles.

I am only summarizing here what was an exhilarating search for the legacy of this illustrious Celtic hero because I have written about it at length in *Adam's Curse*. Soon after this discovery – which was also paralleled by the Macleods of Skye, though there the linkage was to a different chromosome – I heard about the research on Genghis Khan's profligate genetic legacy. His Y-chromosome, passed down through generations of emperor sons, is now found in an estimated 16 million male descendants. This might put Somerled's 200,000 descendants in the shade, but the feeling has grown among geneticists that the Genghis effect could be an important factor in the rise and fall of Y-chromosomes, not only in Asia but in other parts of the world, including the Isles. Recently Brian McEvoy and Dan Bradley from Dublin have found an Irish equivalent to the Macdonalds and the Macleods. Again starting with an unusual Y-chromosome, they noticed its occurrence in a related set of surnames that were linked to branches of the Ui Neill, the clan that had held the High Kingship at Tara, and had expelled the Dál Riata to Argyll. The Ui Neill equivalent of Somerled was Niall Noigiallach, better known as Niall of the Nine Hostages, who lived in the second half of the fourth century AD. This was a time when the Romans were beginning to withdraw from mainland Britain. According to legend, Niall raided and harassed western Britain and specialized in capturing and then ransoming high-ranking hostages, hence his soubriquet. His most famous captive was one Succat, who went on to become St Patrick. Niall's military exploits carried him over the sea to Scotland, where he fought the Picts who were trying to retake the recent Irish colonies of Dalriada. It was during a raid even further afield, in France, that an arrow from the bow of an Irish rival killed Niall on the banks of the River Loire in AD 405.

Niall was succeeded in the High Kingship by his nephew, Dathi, his father's brother's son. This was typical of the Gaelic tradition of *derbhfine*, the rules of inheritance that chose the new king from among the direct male relatives of the old. This served to ensure the patrilineal inheritance of the High Kingship itself and of the whole clan of Ui Neill. Their hold on the High Kingship was remarkably durable, lasting from the seventh to the eleventh century AD. Brian McEvoy's and Dan Bradley's Y-chromosome tests on the Irish showed that a high proportion of men with Ui Neill surnames – names like Gallacher, Boyle, Doherty, O'Connor and even Bradley, as well as O'Neill – shared an identical or very closely related Y-chromosome signature, strongly indicative of direct descent from Niall himself. In the parts of Ireland most strongly associated with the Ui Neill, mainly in the north-west, the proportion of these Y-chromosomes reaches almost one quarter of the male population.

These bursts of Y-chromosome success over a few generations are something to be aware of in our interpretations of the genetic evidence from the Isles. The predictable effect will be to distort the Y-chromosome profile of a region in favour of the local chieftains and also to exaggerate the differences between the regions. We have already seen how this may be happening in that Y-chromosome similarity scores between regions are usually lower than the same comparative score for mitochondria. The only regions that we have so far encountered where the Y-chromosome similarity score is almost as high as the mitochondrial are the two Pictland regions of Tayside and Grampian. If inheritance and succession really were matrilineal, then this practice would indeed neutralize the Genghis effect, since no Y-chromosome could be linked to wealth and power for generation after generation. Another effect will be to reduce the age of a patrilineal clan. If one or a few Y-chromosome signatures come to predominate in a region due to the Genghis effect, they can do so only at the expense of others. These, it follows, are eliminated either because the men who carry them are actually killed, as was the case in the Mongol Empire, or because they do not have their fair share of children, since the Genghis male monopolizes the women in one way or another. The Genghis effect can substantially reduce the variety of Y-chromosomes, so the normal way of estimating the age of a clan in a region by averaging the number of mutations will be distorted. The fewer different Y-chromosomes there are, the fewer mutations will be found, the average will drop, and the age estimate will become artificially younger. The more pronounced the Genghis effect, the greater the distortion and the greater the difference between the true age of a clan and the estimate. To take things to extremes just to illustrate the point, were Genghis Khan's Y-chromosome the only one to have survived from thirteenth-century Mongolia, only the mutations along his line would have accumulated and the age estimates of Mongolian Y-chromosomes would come out around 800 rather than thousands of years.

Scotland has shown us a bit of everything. Vikings, Picts, Celts, the erratic effects of patrilineal kingship and the ancient bedrock of maternal ancestry. We have discovered that the Viking settlement of Orkney and Shetland was very substantial but also much more peaceful than was previously thought, with as many Norse women as men among the settlers. We now know how to identify a Norse Viking genetic presence anywhere in the Isles. Orkney and Shetland aside, there is a very close genetic affinity between Scotland and Ireland. There has certainly been a substantial settlement from Ireland at some time in the recent past, and the Irish Y-chromosome infiltration into the west of Scotland is almost certainly the signal of the relocation of the Dál Riata from Ulster to Argyll in the middle of the first millennium. We have also made an important discovery about the Picts. Their descendants are still in Scotland in force, yet they are not the weird prehistoric relics that were once imagined. They fit very comfortably into the Celtic bedrock of the Isles and they have been here a very long time. The Y-chromosomes are more diverse in the Pictland regions of Grampian and Tayside than in, say, Argyll or Ireland, and the explanation may have something to do with the tradition of matrilineal inheritance. The Western Isles stand out as being a little different from the mainland. There are Viking genes there for sure, but also strange ratios of the maternal clans. The Western Isles have the highest concentration of Katrines in the whole of Scotland, twice that of the Pictish heartland of Grampian, along with large numbers of a maritime branch of the clan of Tara which has travelled from the Mediterranean.

We now have a very good idea of the basic genetic structure of Scotland and Ireland, two of the three regions of the Isles that are most closely identified with a Celtic ancestry. The third is Wales and to reach it we go south down the west coast, through the North Channel and into the Irish Sea. We sail past the Norse outpost of the Isle of Man and head further south to the distant peaks of Snowdonia, the highest point in the land of the Red Dragon.

13

WALES

The smallest in land area of the four regions of the Isles, Wales has a population of just under 3 million, occupying a country of 8,000 square miles. Like Scotland, Wales is a mountainous country, with a broad upland spine running down the centre from the high mountains surrounding the summit of Snowdon (1,085 metres) to the Brecon Beacons, which rise to 886 metres in the south. Between these, the upland plateaux of the Cambrian Mountains are intercut by river valleys radiating in all directions and emptying the abundant rainfall into the Irish Sea and the Bristol Channel. Like Scotland, Wales is bounded on three sides by the sea, backed by coastal lowlands, and shares a land boundary with England. And, also like Scotland, this border, the Welsh Marches, has moved backwards and forwards according to the successes and ambitions of the rulers on either side.

The archaeological evidence for the earliest settlers is comparatively thin on the ground. There is the ancient tooth from the cave at Pontnewydd in north Wales that we met in Chapter 1, but at 300,000 years it is far too old to be from a modern human species. At the other end of the country, at Paviland Cave on the Gower Peninsula, the remarkable burial of the ochre-tinted body of the Red Lady showed that *Homo sapiens* had reached Wales before the last Ice Age, but had been forced to retreat south as the temperature dropped and the great herds left the slowly freezing land.

Precisely when humans returned to Wales is uncertain, but considering how close it is to the Cheddar Caves across the Severn, it is surely likely that Palaeolithic hunters got that far, though they left no trace. There are just a handful of late, coastal Mesolithic sites around the Irish Sea, where conditions were similar to sites in Ireland and on the west coast of Scotland. Life for the early Welsh followed the familiar pattern we have already seen elsewhere in the Isles: a life of gathering shellfish, offshore fishing in coracles and hunting in the wooded slopes behind the seashore. There are Neolithic megaliths in Wales, though none matches the magnificence of the Irish passage graves at Newgrange or the great monuments in Orkney or at Stonehenge. Neither have any neolithic villages like Skara Brae yet been discovered. The earliest houses, round in outline, are isolated. There is some evidence of communal activities around the sites of chambered tombs, the *cromlechs*, whose stark stones have been stripped of their protective mounds of earth.

The history of Wales reads like a catalogue of struggle and resistance against invasion. The Romans were the first to make a serious attempt to subdue the Welsh after the Claudian invasion of Britannia in AD 43. There were, of course, no national boundaries in those days. There was no country called Wales, nor Scotland, nor England. The only defined territories were those occupied by Celtic tribes, in Wales the Silures in the south, the Demetae in the south-west, the Cornovii in mid-Wales, the Deceangli on the north coast and the Ordovices in the mountains of Snowdonia and Cader Idris. Our only knowledge of these tribes and the lands they occupied comes from what the Romans themselves recorded on their campaigns. How accurate this is, we cannot know.

The relentless expansion of the Roman Empire was nowhere near as well thought out as we might imagine. It was much more of a hit-and-miss affair, and when the invasion was launched, the Romans had very little idea of the extent or the geography of Britannia. They probably did not even realize that beyond the easily subdued fertile lowlands lay barren mountain tracts which were far more difficult to conquer and then to hold against stubborn and spirited resistance. The mountains were also hardly worth having anyway, since the land was so poor that it could never yield much in the way of taxes. Everywhere there was the problem of secure frontiers.

The first frontier in western Britannia between the Romans and the unconquered tribes followed the diagonal course of the Fosse Way from Exeter to Lincoln. This proved to be an unstable border and was repeatedly attacked by the Silures in AD 47 and 48, encouraged by Caratacus, the fugitive chieftain of the defeated Catuvellauni who had taken refuge in Wales. To contain the Silures, the Romans built fortresses at Gloucester and Usk. Caratacus moved north to the Ordovices in Snowdonia, and after their defeat in AD 51, and the capture of his wife and children, he fled to the court of Queen Cartimandua, leader of the Brigantes in northern Britannia. There his flight ended and he was handed over to the Romans by Cartimandua and taken to Rome in chains. Rather than execution, which captured 'rebels' could usually expect,

Caratacus was released by Claudius after a defiant speech in which he is said to have exclaimed, referring to the grandeur of Rome, 'Why do you, who possess so many palaces, covet our poor tents?'

Welsh resistance to Rome did not end with the capture of Caratacus. The Silures resumed their attacks and defeated the twentieth legion in AD 52. Eventually the Emperor Nero, who had succeeded Claudius in AD 54, issued instructions to subdue the entire island of Britannia and in AD 58 a new governor arrived in Britannia to carry out the orders of the Emperor. Suetonius Paulinus was a professional soldier with campaign experience in the Atlas Mountains of Algeria, so he was used to dealing with independent-minded tribesmen the hard way. In two years of campaigning he had Wales in an iron grip. Refugees fled to Anglesey, the centre of the Druids, and Suetonius launched an attack. Tacitus records the scene with the British lining the shore of the Menai Straits: 'Among them were black robed women with dishevelled hair like the Furies, brandishing torches. Close by stood the Druids raising their hands to the heavens and scream-ing dreadful curses'.

Tacitus was a historian, but he needed to sell books, so his popular histories were always written to appeal to his readers in Rome. His description of the Druids' habit of 'drenching their altars in the blood of prisoners and consult[ing] their gods by means of human entrails' was bound to boost sales.

Neither the sight of wailing women nor the threat of eviscer-ation on a Druid altar was likely to deter Suetonius Paulinus. His troops swam across the Straits and easily defeated the Celtic refugees and the Druids, not only killing all they could find but destroying the groves of trees that were sacred to their religion. However, Suetonius was forced to withdraw immediately to deal with the revolt of the Iceni under Boudicca and it was left to

Tacitus's father-in-law, the general Julius Agricola, to complete the subjugation of the Ordovices in AD 78, which he did, according to Tacitus, by killing them all, before moving off the following year to deal with Caledonia.

Containing the Celtic tribes of Wales proved to be a long and costly operation for the Romans. Legionary forts at Chester, at Wroxeter near Shrewsbury and at Caerleon near Newport in south Wales defined the boundary between the rebellious uplands and the subjugated lowlands. Smaller forts at Caernarfon in the north-west and Carmarthen in the south-west contained Wales within a fortified rectangle, supplemented with a network of camps and smaller forts placed one day's march apart and connected by straight roads. The military presence was strongest in the lands of the most belligerent tribes, the Ordovices in Snowdonia – so some must have survived Agricola – and the Silures of the south. The other Welsh tribes, the Deceangli along the north-western coastal plain between Conway and Chester, and the Demetae of Dyfed, showed less appetite for resistance and their territories were accordingly less densely garrisoned. Eventually the Celtic tribes of Wales settled for the life of a distant outpost of the Empire. The Romans took gold from Dolaucothi in mid-Wales back to Rome to be minted into coins and mined copper from the Great Orme near Llandudno. The Romans began to withdraw their garrisons from Wales by the beginning of the second century, indicating that the inhabitants were coming to terms with the Roman occupation, the last to succumb being the Ordovices.

What might have been the genetic consequences of the Roman occupation that we should look out for? After the initial campaigns of subjugation, which may well have resulted in the deaths of thousands of men, the military outposts became important centres of economic activity. Around Caerleon, for example, a small township or *vicus* grew up outside the walls of the fort. By AD 100 there were 2,000 people living in the Caerleon *vicus*, attracted from far and wide by, and dependent on, the great wealth, in comparative terms, of the garrison. Even though there were rules which banned official Roman marriage between the legionaries and the indigenous people before AD 190, unofficial liaisons were tolerated. Indeed, as the threat level fell, garrisons were reduced in size and troops were withdrawn to be redeployed elsewhere in Britannia; this had a severe effect on the economy of the *vici*. And not only on the economy, according to one historian, who points out the effect that the redeployment of the garrison would have had on the women who had borne children. They had to stay behind.

As usual, if there is one, it will be the Y-chromosome that is the witness to this activity. But who were the soldiers of the Roman army? Not all from Rome, that's for sure. After the initial campaigns, when there would have been a substantial Italian contingent in the legions, the occupation itself was left in the hands of the auxiliaries. In Wales these troops, who would be granted citizenship when they retired, were drawn largely from the valleys of the Rhine and the Danube. It is for Y-chromosomes from that part of Europe that we should keep an eye out as a sign of the genetic influence of the Roman occupation.

After the withdrawal of the Roman army from Wales in the fourth and fifth centuries AD, the demilitarized population came under attack from the Irish, including the infamous Niall of the Nine Hostages. In a mix of raiding for slaves and settlement, reminiscent of the first decades of the Viking age in Scotland, the coast of Wales facing the Irish Sea endured continual attacks. This period of attempted Irish colonization coincides with the expansion of the Dál Riata into Argyll, only 100 miles to the north. It may even have been carried out by the same people, and for the same reasons: the ambitions of the Ui Neill. But the Irish never established themselves in Wales as successfully as they did in Argyll. There was no equivalent in Wales of the continuous friction in Scotland between the Picts and the Gaels of Dalriada. The Irish form of Gaelic never displaced the P-Celtic of the Welsh as it did in Scotland.

Within Wales, the people divided into a succession of minor kingdoms and before long the disputed land frontier became a battle zone once again, as it had been during the first years of the Roman occupation. This time the enemy were the Saxons, who had arrived in England in the middle of the fifth century and who, like the Irish, took advantage of the power vacuum left behind when the Romans departed. There is more to come on the Saxons and their genetic legacy when we travel to England, but for the time being we need only know that their westward expansion was effectively halted at roughly the same frontier that the Romans had defined with their lines of legionary forts.

The boundary was formally marked out in the late eighth century by Offa's Dyke, named after the Mercian king responsible for its construction. Unlike Hadrian's Wall, Offa's Dyke was not a fortified frontier barrier with regularly spaced garrisoned forts, but an earthwork built to denote rather than to defend the frontier, though in its construction it was far more than a boundary fence. Offa's Dyke consisted of an earth embankment up to 3 metres high and backed by a ditch up to 20 metres wide. The boundary it defines stretched for 240 kilometres from Prestatyn on the north coast to Beachley near Chepstow on the Severn Estuary. The Dyke marks this boundary for 130 kilometres, the rest being defined by natural features like the

River Severn. Though it is built only of earth, thousands of men must have been involved in its construction, proof of the level of organization in the kingdom of Mercia at the time.

The Saxons did not advance far beyond the Dyke but, as you might by now expect, it proved to be a fluid boundary. Though the construction of the Dyke coincided with the beginning of the Viking Age, the Welsh kings did not respond by uniting under one leader as the Celts and Picts had done in Scotland. The Welsh never did regain the lost lands in England on behalf of the Britons, though not always through want of trying. In 633 Cadwallon launched a counterattack against the Saxon King Edwin, whose title *Bretwalda* at least claimed control of the whole of Britain. Edwin had attacked Anglesey, but Cadwallon drove him back into England and eventually defeated and killed him at the battle of Meigen near Doncaster. He then killed Edwin's heirs, Osric and Eanfrith, and, according to Bede, it was his intention to exterminate the whole English race. He had his best and only chance in 633 for, the following year, he was himself killed by Eanfrith's brother. As we shall see, the memory of Cadwallon's near success was to shape things to come.

We have seen what a significant genetic effect the Viking settlements from the late eighth century onwards have had in Scotland. Can we expect the same in Wales? Although the Vikings soon dominated the western seaways and had, by 830, begun to set up colonies at Dublin and other Irish coastal towns, there is very little evidence of them having succeeded in colonizing Wales. In the north they were actively repelled by Rhodri Mawr (Rhodri the Great), King of Gwynedd, who defeated a Danish attack on Anglesey in 856.

Only in the far south-west is there any suggestion of Viking settlement. It is there, as we saw in an earlier chapter, that the high levels of blood group A have been used to argue for a substantial Viking settlement in what is now Pembrokeshire. We shall certainly see if we can find corroborative evidence when we look at the genetics. Based on the experience in the Northern Isles, if Viking genes are there in large numbers we will certainly find them.

The Welsh kings continued in their internecine wars, sometimes making alliances with the Saxon kings against one another. So long as they were busy fighting between themselves, they were no threat to England. Only once did they unite under a single ruler, and then only for six years. Gruffudd ap Llywelyn began as the King of Gwynedd and it was from this position that he launched a campaign of murder and usurpation against the other kings that culminated in his recognition as the King of all Wales by 1057. Gruffudd's campaigns against Mercia on the border with England revived the memories of Cadwallon, the last Welsh king to interfere in English affairs, so in 1063 the English decided to do something about it. Harold, Earl of Wessex, went after him. Gruffudd was pursued back to Snowdonia, where he was killed by the son of one of his royal victims. To show there were no hard feelings, Harold married Gruffudd's widow, Ealdgyth, the granddaughter of Lady Godiva. When Harold became king in January 1066, Ealdgyth became Queen of England after six years as the first, and only, Queen of Wales. Her reign as Queen of England was even shorter: it came to an abrupt end when Harold was himself killed by the Normans at the Battle of Hastings the following October.

The Norman Conquest had immense repercussions for life in England almost immediately. For Wales, the old border held the Normans at bay – for a while. Compared to Scotland, with its 200 years of unified rule under the descendants of Kenneth

MacAlpin, Wales was in a shambles after the downfall of Gruffudd ap Llywelyn. Feuds between claimants to the now vacant kingdoms had created a chaos of murder and betrayal which culminated in the battle of Mynydd Carn in 1081. Two royal houses emerged: Gwynedd in the north and Deheubarth in the south.

Again the perpetual problem of a secure border with Wales presented itself to the first Norman king, William the Conqueror, just as it had to the Romans and the Saxons before him. He had no interest in the conquest of Wales, but he did want a stable frontier. His solution was to grant lands along the frontier to his most reliable barons and, without positively encouraging them, to turn a blind eye if they felt like expanding their holdings into Wales. These men, the Marcher Lords, began by building castles along the frontier, first of earth and timber, then of stone. Then they really let rip and spilled over the border in deadly earnest.

By 1093 the most aggressive of the Marcher Lords, the Earl of Shrewsbury, reached the Irish Sea coast at Cardigan at the mouth of the River Teifi. Up went a castle. From there he pushed south into Dyfed and built the huge castle at Pembroke. Another Marcher Lord launched an attack against Rhys ap Tewdwr, the ruler of Deheubarth, who was killed at Brecon in 1088 resisting the advance. It was the death of Rhys ap Tewdwr that, to later historians, marked the final demise of the Welsh kingship. It looked as if nothing could save the Welsh from the Norman threat. However, the Welsh did manage to fight back. The forces of the Marcher Lords were expelled from Gwynedd, Ceredigion around Cardigan and from most of mid-Wales, but they hung on around Pembroke, Glamorgan and Brecon. The Norman domination was never complete and there was a resurgence in the position of the Welsh princes. In the Treaty of Montgomery in

1267, Henry III recognized Llywelyn ap Gruffud as the first 'Prince of Wales' with control over several of the old Welsh kingdoms.

However, Henry's successor, Edward I, decided to conquer Wales once and for all and in 1277 led his army of 800 knights and 15,000 infantry into the heartland of Gwynedd, stronghold of Llywelyn, and forced his submission. Edward continued his campaign through Wales, building a new series of castles, including the impregnable structures at Conway, Harlech, Beaumaris and Caernarfon. A revolt in 1282 gave Edward the excuse for another campaign. This time the Welsh fared better and defeated Edward's army on more than one occasion. However, Llywelyn himself was killed near Builth in December 1282 and resistance had collapsed by the following summer. In 1284 the Statute of Rhuddlan set out England's sovereignty over Wales and in 1301 Edward's son, who became Edward II, was invested with Llywelyn's title 'Prince of Wales' at an elaborate ceremony at Caernarfon Castle. With the exception of Edward II himself, every subsequent British monarch has given the title 'Prince of Wales' to their eldest son.

The Welsh made one final attempt to free themselves from English domination. In 1400, taking advantage of the confusion caused by the overthrow of Richard II, the Welsh rose up in revolt under Owain ap Gruffydd Glyn Dwr of Glydyfrdwy, better known outside Wales by the English translation Owen Glendower. On 16 September 1400 he was proclaimed Prince of Wales at Bala and his followers began their quest to regain the independence of Wales by attacking nearby English settlements at Ruthin. Intriguingly, Owen Glendower used his alleged descent from the legendary Brutus, first King of the British, to back his claim. He reigned for twelve years, even convening a

Welsh parliament at Machynlleth in mid-Wales and he was recognized as sovereign of an independent country by the King of France. The revolt was eventually ended by England's military superiority. Many of the great castles built 100 years earlier by Edward I had never surrendered, and by 1414 the army of Glendower surrendered at Bala. Owen Glendower himself was never captured and, rather like his 'ancestor' King Arthur, he vanished into the mists. Finally, in 1563, the Act of Union formally combined the political fortunes of England and Wales.

14

THE DNA OF WALES

Wales is the only part of mainland Britain where the original language is still spoken. We might take that as an indication that there has been very little disturbance of ancient Welsh culture, and maybe very little disturbance of the indigenous genetic make-up. But it is clear from Welsh history that there have been very many foreign intrusions on to Welsh soil from the Roman period onwards. What we do not know is the magnitude of their genetic effect. Traces of Viking DNA are a strong possibility in Pembroke, and the effects of the Saxon and Norman incursions may have had substantial genetic consequences.

Our campaign in Wales, for that is how it seemed, began in the early days of the Genetic Atlas Project. Four of us set off by car in a planned series of swoops on secondary schools throughout the Principality. This was in the days before we had discovered the easy delights of the DNA brushes. We needed blood. But we had not arranged to visit blood-donor sessions in Wales. We had yet to refine that approach. The blood samples we used in the early days were taken from fingerpricks, the collection of which had unintended consequences. One of my research team, Kate Smalley, had once been a teacher and she realized that hard-pressed sixth-form Biology teachers might welcome a visit from outside scientists if we gave a lecture and, in return, we may be able to ask for volunteers. That would give the teacher a double period off, if nothing else. We chose Oswestry, a market town on the English–Welsh border not far from Shrewsbury, as our first destination. My main concern was that, however well the lecture on our project went down, it might be hard to get volunteers to submit to a fingerprick blood test. The automatic lancets, the ones diabetics use to take a sample for blood-sugar measurements, drive a short needle into the skin. It isn't painful, but neither is it completely painless.

I had been through my presentation and the time to ask for volunteers had arrived. I was met by a sea of blank faces. 'It really doesn't hurt,' I entreated. There was no reaction. I suddenly realized what I needed to do. I got out a lancet and pulled back the spring-loaded trigger. I wiped the tip of my left index finger with an alcohol swab to sterilize it and, 'ping', lanced it, trying not to wince. That did the trick and soon we had everybody lancing their own fingers or, better still, their friends'. The drops of blood were soaked up on special cards, which we knew would keep the DNA safe until we got back to the lab. I don't think we would be allowed to take blood these days. Everyone is so scared of it.

After Oswestry we divided into two teams of two, one heading north to Anglesey while Kate and I set out for Bala and Dolgellau. At Bala I discovered the unexpected advantages of the fingerprick technique as a way of collecting DNA in schools. It is this. Because there is some discomfort involved, to take the test is a mini-act of bravery. And once the children had done it, what better thing to do, at the break after the lesson was over, than to run to their friends in other classrooms and taunt them into having the test. It certainly worked. Once they had given the blood sample, the children were running off round the school, to the staffroom and the canteen, collecting more volunteers. They had started a chain reaction. A queue of children, teachers and dinner ladies formed and we were busy for at least another hour. By the end we had over 200 samples from Bala, practically the entire school. By the end of the week, we had been to twenty schools and collected over 2,500 samples. Fantastic.

What are we on the lookout for in Wales? The early blood-group work, as well as proposing the Viking settlers/Flemish weavers solution to the elevated blood group A frequency in Pembrokeshire, also noted very high levels of blood group B around the Black Mountains at the western end of the Brecon Beacons south of Llandovery. There is also abundant work from the early twentieth century by H. J. Fleure, an eminent anthropologist based at Aberystwyth University, on the unusual head shapes and Neanderthal-like faces of people living in the remote mountains near Plynlimmon in mid-Wales at the headwaters of the River Severn and the River Wye.

Plynlimmon is not very far from the market town of Tregaron, where, while staying at the Talbot Inn in the market square one October night on another visit to collect DNA samples, I was told the fantastic story of the Tregaron Neanderthals. The Talbot Inn is an old drovers' inn dating from the thirteenth century, complete with stone walls, oak beams and open fires. It was a dark night and the rain had not stopped all day. The fire was blazing away and there were a few local men at the bar, staring at their pints of bitter and glad to be out of the rain. We got talking, and before long I was telling them about what I was doing in that part of Wales and about the Genetic Atlas Project. We had evidently been overheard by a man sitting alone at a small table. He beckoned me over and I sat down. And then he began to tell me about the elderly twin brothers, both bachelors, who had lived at the end of a long track leading into the Cambrian Mountains behind the ruins of the Cistercian monastery at Strata Florida, further up the Teifi from Tregaron. I knew this track, as once in my youth I had been up it looking for an incredibly rare bird, the Red Kite. Now, thanks to successful reintroductions to the Chilterns, anyone can see these beautiful birds gliding and twisting every time they travel on the motorway between Oxford and London. But, back then, there were only a few pairs left, all of them in mid-Wales. I had heard that a pair was nesting in the woods behind Strata Florida and I remember walking for several miles up into the hills, first through the woods then up on to the grassy uplands. I did not see a Red Kite, but I do remember seeing a cottage, up a side track, which, from the washing on the clothes line, was clearly inhabited. I think this must have been the place. I don't remember any other dwellings.

My companion at the Talbot told me that the men who lived in this cottage in the 1950s and 1960s were Neanderthals. This fact was well known. So well known that a visit to the brothers was on the history syllabus at Tregaron school. Every year, in the summer term, the third-form History class would take the school van as far as they could up the track and the children would walk the rest of the way to the cottage. The Neanderthals obviously looked forward to the visits because, on the appointed day, they made sure they had plenty of cakes and lemonade. The children stayed for an hour while the teacher explained about human evolution and where the Neanderthals fitted into the scheme of things. Then they left and walked back down the hill to the van.

Of course I didn't actually believe these men were Neanderthals any more than I am. But I do still hope one day to find just one person with Neanderthal DNA. It is a vanishing hope as more and more DNA is tested from around the world. But could I recognize it if I found it, whether around Tregaron or Cardiff, or London or California? The answer is definitely yes – so long as it is mitochondrial DNA.

I had once attempted, but failed, to recover Neanderthal DNA from the Tabun skull from the Natural History Museum in London. The Tabun skull was dated to 100,000 years ago and the teeth looked in fairly good shape. But when I tried to drill into a molar tooth, it was rock hard and I was terrified it would fracture. I did get a little dentine powder from the inside, but I did not smell the reassuring scent of burning flesh, the smell that meant success. However, I did manage to recover a few molecules of DNA from the Tabun tooth. When I put them through the DNA analyser, the mDNA sequences looked distinctly modern, with their closest matches in Israel, where the skull had been excavated. The big debate at the time, in the early 1990s, was whether Neanderthals were an extinct species of human, in which case their DNA should be very different from ours, or whether they were just a phase in the evolution of modern humans, in which case the DNA should be reasonably similar. I never felt confident enough about proclaiming that the modern-looking DNA that I had recovered from the Tabun skull was really from the skull, rather than from the archaeologists and museum curators who had handled it over the fifty years since it was excavated.

I am glad I was cautious, because two years later what did appear to be genuinely ancient DNA was recovered from the

Neanderthal type-specimen, the original one that had been found in the Neander Valley in Germany (*Tal* is valley in German) in 1863. This DNA was very different from any modern DNA. It had 27 mutations in comparison to the mitochondrial reference sequence, while even the most distinct modern DNA only varies from the reference by 12 changes. When similar DNA was found in two further Neanderthal remains, from Croatia and the Caucasus mountains, it provided reasonable proof that Neanderthals were indeed an extinct species of human. The last Neanderthal died in southern Spain about 27,000 years ago; at least that is where the most recently dated remains have been found. But that was before the world knew about the Tregaron twins!

The brothers had passed away in the 1980s, so another trip up the track into the hills would be pointless. Since they were men, and bachelors at that, their mitochondrial DNA could not have been passed on to their children, even if they had any. And neither the man at the Talbot Inn, nor anyone else I spoke to in Tregaron, knew where the brothers had come from, so I could not track down a relative. The only chance was that, among the smiling children at the local school, there was one who, through maternal connections, would carry the tell-tale Neanderthal DNA. There was a lot to look out for in Wales.

Examining first the matrilineal DNA from Wales, the living record of the journeys of women to this part of the Isles, the pattern of maternal clans is very similar to Ireland, and to what we have also seen in the two Pictland regions of Scotland, Tayside and Grampian. The clan of Helena predominates, as always, with 47 per cent of people in both regions belonging to that clan. When Ireland is compared to the whole of Wales, this close similarity extends to the other clans as well. When I divided Wales into three regions, north, mid- and south Wales, a few differences did emerge, mostly ones that showed a closer genetic link between north and mid-Wales than either did to the south of the country. But the overall pattern was one of continuity with Ireland and, to a lesser extent, with the Pictland regions of Scotland. But, unfortunately, there was no sign of any Neanderthal mDNA.

When I looked at the patrilineal Y-chromosomes in the three regions of Wales, the pattern was extremely interesting. There were two outstanding features. First, there was practically no sign of Norse Viking settlement. If you recall from the Northern Isles and from Norway itself, there is a high concentration of members of Sigurd's clan; 20 per cent of Shetland men are in this clan. And yet in Wales there are virtually no men from Sigurd's clan. I interpret this as strong evidence against any substantial Norse Viking settlement in Wales. The only hint of Viking ancestry is in the north, where just three men, Mr Roberts from Bangor, Mr Owen from Llanfair and Mr Davies from Meifod, are in the clan of Sigurd. At such low frequencies we must doubt whether they have inherited their Sigurd chromosomes from Vikings directly or in their transmuted form in the blood of a Norman. Since there were no Sigurds at all in our samples from south Wales, which was far more heavily occupied by Normans than was the north, I tend to think that these three gentlemen are more likely to be of direct Viking than Norman ancestry. You will recall that there were Viking raids on Anglesey which were actively repelled by Rhodri Mawr in 856. Perhaps it was from action around this time that Messrs Roberts, Owen and Davies acquired their Viking ancestors. Their detailed fingerprints are certainly matched in Norway.

There was just one Sigurd in mid-Wales, Mr Jones, from the small village of Garthmyl near Rhyader. And none at all in south

Wales, even in Pembrokeshire where the high level of blood group A was explained by a Viking settlement in the area. There would need to have been a very large influx of Vikings into Pembrokeshire to alter the blood-group proportions of the whole region and we would have been bound to find several Sigurds in the vicinity. But we did not find a single one. I think that has to mean that the Viking explanation of Morgan Watkin for the high frequency of blood group A in 'Little England beyond Wales' is wrong.

Turning to the clan of Wodan, this hovers around the 10 per cent mark in all three regions of Wales. However, when I looked at the detailed fingerprints, I found a small cluster in mid-Wales that caught my eye. There were only half a dozen of them, but they were unusual. Mr Rees from New Quay, a picturesque fishing port on Cardigan Bay, Mr Jones from Mynachlog near Tregaron, and finally Mr Davies from Lampeter.

Before I draw any profound conclusions, may I recommend Lampeter as the best place in Wales for ice-cream. At the junction of the High Street and the Tregaron Road stands the ice-cream emporium of Conti's Café. Going inside, when I was last there, was like returning to the cafés of my youth. No cappuccinos or lattes here, just weak milky coffee in one of those unbreakable glass cups, served by a waitress in a blue tabard. A rare experience indeed these days. Alas, I've heard that the interior has been recently revamped, but the ice-cream is still wonderful. Made every day on the premises by the owner, Leno Conti, not brought in ready-made. Perish the thought.

Now for the profound conclusions. I think this Wodan Y-chromosome has been in mid-Wales for a very long time. There are first-generation derivatives nearby, by which I mean Y-chromosomes that have diverged away by one mutation. And it is only one mutational step removed from a chromosome cluster in Pictland. I have not found this chromosome in Ireland or in England, except in one place. Mr Roach, from Sidmouth in Devon, has it. I could be wrong, but I don't think this is a Norman chromosome. If it were, I would have expected to find similar chromosomes in other parts of England, which, with the exception of Mr Roach, I have not. I couldn't help wondering if this is a very ancient Welsh chromosome. After all, Tregaron and Lampeter are not that far from Plynlimmon where H. J. Fleure was convinced from his work on skull shapes that he had found a relic population, and where there was also a very high frequency of blood group B. I wonder, as I write this, whether the great anthropologist ever tasted Conti's ice-cream on his travels.

Of course, we must not forget the clan of Oisin. This is far and away the most common clan in all the three regions of Wales, which it also is in the whole of the Isles. In fact, at 86 per cent, mid-Wales has the highest proportion of Oisin in the Isles outside Ireland. Interestingly, the Pictland region of Grampian is only just behind, with 84 per cent. Only Munster and Connacht in the west of Ireland have higher proportions of Oisin. The Atlantis chromosome, the prevalent Y-chromosome in the clan, is very frequent in Wales, more so even than in Ireland, as a proportion of Oisins as a whole.

There is one other interesting thing to point out. The diversity, that is the variety, of different Oisin Y-chromosomes is lower in Wales, especially mid-Wales, than anywhere else in mainland Britain. Geneticists usually put that down to a recent arrival date, there having been less time for mutations and diversity to have arisen. But to find the lowest diversity in mid-Wales of all places seems very peculiar to me, since all the other historical indicators suggest that mid-Wales has been among the most stable and longest settled of any region in the Isles – even if I did not find any evidence of Neanderthals. The lower than expected amount of accumulated mutations in the Y-chromosomes is beginning to be a recurrent feature of most of the Celtic regions of the Isles. Whether this is also true of England, we are about to discover as we push east over the hills.

15

ENGLAND

As we cross the long-disputed boundary into England, the land spreads out in all directions, undulating certainly but without the mountains that insulated Wales, and Scotland, against the full force of foreign invasion which began with the Romans and continued for more than 1,000 years. Geography, as always, led history by the hand. It was the fertile lowlands of England, not the barren hills of Wales or Scotland, that made the Isles such a tempting target from the Roman invasion of AD 43 to the Norman Conquest of 1066 and beyond. But the settlement of England and the Isles began thousands of years earlier.

England is home to 49 million people, which is almost 80 per cent of the entire population of the Isles, packed into 50,000 square miles, which is 40 per cent of the space. England has examples of almost every kind of geological structure, from extremely old volcanic rocks in Cornwall and Cumbria to very recent, reclaimed soils in the fenlands of East Anglia. Between these extremes of age and distance lie successions of sandstones and limestones from different geological eras which cross the country diagonally from the south-west to the north-east. As a rule of thumb, the further east, the younger in geological time the rocks become. These sedimentary bedrocks, built up over hundreds of millions of years when England lay beneath a warm and shallow sea, are mainly alkaline. They erode to very fertile soils, and almost all of England is now intensively farmed. It was always the agricultural wealth of England and the opportunities this provided for taxation and tribute, as well as settlement, that attracted the attention of foreign invaders.

Beyond the fertile plains and rolling downland, England is surrounded by mountains and high hills. Along the centre, the spine of the Pennines in the north of England rises to 893 metres at Cross Fell. Forty miles to the west, among the picturesque mountains of the Lake District, is Scafell Pike (977 metres), England's highest peak. The garland of mountains and hills which form the boundaries with Wales and Scotland have pre-occupied all invaders as they tried, and usually failed, to protect England behind stable frontiers.

The first people to reach England after the Ice Age were the big-game hunters of the Old Stone Age, colonizing the Isles directly over the land bridge from continental Europe. By 12,000 years ago, hunters were living in the caves at Cheddar. After the cold snap of the Younger Dryas forced a temporary retreat, the Mesolithics returned 10,000 years ago. They were confined to the coasts and riverbanks by the dense woodland that soon covered the warming Isles. Like the occupants of Mount Sandel in Ireland and Oronsay off the coast of Scotland, they were semi-nomadic, with winter and summer camps alternating between woodland and shore to make the most of the wild food: fish and shellfish in winter, birds' eggs in spring and summer, hazel and other nuts in the autumn – and red deer at any time they could be killed. The Mesolithic life in England was no different from in the rest of the Isles, and is nowhere more completely documented than at Starr Carr in the Vale of Pickering in North Yorkshire, 5 miles to the west of the seaside resort of Scarborough.

At the marshy edge of a lake, this was a site where, 9,500 years ago, the elusive Mesolithics brought their kills from the nearby high ground of the North Yorkshire Moors to be butchered and distributed. Thanks to the marshy, waterlogged conditions, all sorts of things have been preserved which on dry sites would have been lost. Pollen, insects, charcoal, wood and animal bones are all preserved in the damp and airless peat. From an analysis of the bones left at Starr Carr, most of the meat came from wild cattle, the enormous aurochs which roamed through the dense woods. There were elk and red-deer bones too, sometimes with marks to show where a flint-tipped arrow had cut through the skin on its way to the beast's heart. Badger, red-fox and pine-marten bones show that even smaller mammals could be killed, perhaps for food, perhaps only for their skins. The Mesolithic occupants of Starr Carr were extremely skilled in working deer antler, making not only large objects like spearheads, but also smaller, but still deadly, arrowheads. The flints they used to work the antlers lie all around the site. But perhaps the most remarkable revelation at Starr Carr is the evidence of domestic dogs. The hunters, we can assume, used these dogs to round up deer and wild cattle or to pursue a wounded animal if an arrow had failed to find the heart.

Farming arrived in England a little before it did in the rest of the Isles. Bit by bit the wild woods were cleared. The Mesolithics already knew how to kill trees by ring-barking, so they created glades to encourage the growth of hazel bushes. The Neolithic farmers killed trees in the same way, and may have been descendants of the same people. They targeted elms in particular, because they understood that they grew in the most fertile soils. Gradually, more food was grown than was strictly necessary for survival, which meant that not everyone had to spend all their time looking for food. Thus began the social revolution that culminated in the rise of chieftains and then minor kings, each battling it out for supremacy and ownership of land. Megalithic monuments, like the stone circles at Stonehenge and Avebury, took pride of place in a landscape rich in burials and tombs. The newly discovered metals of copper, bronze and iron, in that order, replaced bloodstone and flint as the principal materials for axes, knives and other agricultural implements. They also found their uses as weapons, cast or beaten into daggers, swords and spears as warfare became endemic. Iron tools, much stronger than bronze and with a much sharper cutting edge, made woodland clearing easier. The increase in the acreage of agricultural land led to a big rise in the population of the Isles.

By the fourth century BC, the archaeological evidence points to an increase in inter-tribal warfare. Hill forts became more numerous and their defences more elaborate. Swords replaced daggers in a sign of more organized fighting. By the third century BC, the style of metal-working for both weapons and jewellery had changed to the second Celtic phase of La Tène, but always with a distinctive British dialect. The export of Cornish tin, an essential ingredient in the manufacture of bronze, continued apace, with the export trade to the Mediterranean dominated by Phoenicians.

One of the very first accounts of the Isles of the time, by Pytheas from the Greek colony of Massilia (now Marseilles) in southern France, was written around 320 BC. His original work, *On the*

Ocean, has not survived and we only know of his remarkable journey through references to it from other classical writers like Eratosthenes and Pliny. Pytheas probably travelled overland from Massilia to the mouth of the Gironde, near present-day Bordeaux, and boarded a ship bound for the north. It took him three days to sail up the coast of France and around the edge of Brittany. From there, his journey took him across to Cornwall and the *Prettanic Isles*, as he calls Britain. He noted the lengthening day as his voyages took him right up the eastern side of Britain to the Orkneys. From there he travelled even further north to a land of frozen seas and volcanoes. This must have been Iceland, though whether he actually went that far north himself or only sailed as far as the Shetlands and recorded the tales of sailors he met there is still keenly contested among historians. *On the Ocean* is important in two ways. It brought the Isles to the attention of the classical world, and it also showed how active were the sea lanes up and down the Atlantic coast of France and all round the British coastline as well. Pytheas seemed able to pick up a sea passage whenever he wanted one.

The need to impress and confirm status with material objects was at least as prevalent then as it is now. The desire for displays of wealth led to the creation of astonishingly beautiful objects and ceremonial weapons. The finds from the royal burial at Sutton Hoo near Woodbridge in Suffolk, from the seventh century AD, in the middle of what we now refer to as the Dark Ages, are delicate and beautiful beyond belief. This was almost certainly the tomb of King Raedwald, a Saxon king from the 620s. Buckles and strap-mounts of gold inlaid with garnets and *millefiori* glass, so fresh and so delicate that, when I saw them on display in the British Museum, it was very difficult for me to believe that they were the originals and not modern copies. The ceremonial shield, the inlaid helmet – these were not objects to be used in battle; they were strictly for display only. Even the sword, its blade forged from eighteen laminated iron rods twisted together and beaten flat, was purely for show. There the display is a modern replica. The original lies with its iron blade rusted, peeled and pitted. But the handle ends with a gold and garnet cloisonné pommel as bright and fresh as new.

In the centuries preceding the Roman conquest, life in the Isles followed the progression widely found across continental Europe. Iron replaced bronze as the principal metal. Fortified encampments developed on the hills. Although there were still extensive forests, much of the land had been cleared for grain or pasture. In the centuries before Caesar's expeditions in 55 and 54 BC, the Isles, and England in particular, had adopted many of the artistic styles of the continental Iron Age. The perennial question as to whether these cultural changes were the consequence of large-scale immigration or of the indigenous people copying and adapting new styles has never yet been confidently answered – and is one that genetics should be in a better position to explore than most disciplines.

Caesar's expeditions set the pattern for the Roman invasion proper a century later. What prevented Caesar himself from embarking on a full-scale invasion, or even if this was his intention, is not known. Certainly he had his hands full in controlling rebellions in Gaul and his ambitions may have been curtailed by such practical considerations. Nevertheless, his expeditions set the pattern for the later invasion. Caesar had forced the surrender and submission of tribal leaders in Britain and had exacted annual tribute payments from them. He also installed puppet kings. So, although there was no permanent occupation, the political influence of Rome was already substantial well before the invasion proper. The British aristocracy began to adopt the trappings of Roman civilization, particularly in the south-east where there was vigorous trade with the nearest parts of Gaul. Britain was exporting corn, iron and cattle to the Roman Empire across the busy sea routes to the ports of Gaul, while Roman luxury goods flowed in the opposite direction. Even if Britain was not part of the Empire, it certainly benefited from the proximity and the requirements of its armies.

The full integration of Britannia into the Empire was only a matter of time. Under Caesar's successor Augustus, and even under Tiberius who came after him, there was no appetite for invasion, even though it would have been comparatively easy. But the taxes were flowing in and Britain posed no military threat. A few troublesome Gauls might have crossed the Channel to escape the wrath of Rome, but that was all. One British tribe, the Catuvellauni, centred on Hertfordshire, began to expand their territories into the lands of neighbours who had thought they enjoyed Rome's protection. But the Romans, now under Augustus, turned a blind eye to these infringements, enabling Cunobelinus, King of the Catuvellauni, to move his headquarters to Colchester, the former base of the Trinovantes, from where he could control the trade routes across the North Sea to the Rhine.

In a re-run of the age-old story, a disgruntled prince – in this case it was Amminius, one of the sons of Cunobelinus – fled to the emperor for assistance. By now the emperor was the notoriously unstable Caligula, who claimed that by accepting the formal submission of Amminius he had actually negotiated the surrender of the whole of Britain, and he issued orders for an invasion to consolidate the surrender. That was abandoned at the last minute, but only after Caligula had reached the Channel coast with his armies. He collected some sea shells and ordered a withdrawal back to Rome.

Although this was a farce, all the ground work had been done. The military build-up, the logistics of invasion, the public relations with the citizens of Rome: everything was in place, so it was an easy matter for Caligula's successor – after his welcome murder – to give the signal to invade. The new emperor was Claudius, Caligula's uncle. Widely thought of at the time as mentally retarded, he was nothing of the sort. Claudius needed a military triumph to cement his authority and Britain was the obvious target. The excuse was an invitation from Verica, King of the Atrebates, who had been expelled following an internal palace coup. The invasion force that assembled on the Channel shore comprised four legions: the II Augusta and XIV Gemina from the upper Rhine, the XX Valeria from the lower Rhine and the IX Hispania from Pannonia in modern Hungary, each with about 5,000 men and an equal number of auxiliaries. The legionnaires were all Roman citizens, mainly drawn from Italy at this period, while the auxiliaries were recruited from native fighters from previously conquered regions of the Empire and organized into regular regiments with Roman commanders. Forty thousand men in 600 ships, under the command of Aulus Plautius, who had seen service in the Balkans, crossed the Channel from Boulogne in Gaul to land on the shingle at Richborough, near Sandwich on the east coast of Kent.

The landings were unopposed and, after digging defensive ditches at Richborough, the troops advanced rapidly to the River Medway, 20 miles to the west, where the British defence under Caratacus and Togodumnus, joint leaders of the Catuvellauni after their father Cunobelinus's death, lay in wait. The British assumed that a major river crossing would deter the advancing army. But Paulinus sent across a contingent of Batavian auxiliaries who were trained in swimming across rivers in full armour. The Britons wore little or no body protection and their long, slashing swords were no match for the short, stabbing *gladius* of the Romans in close combat. Unable to halt the Roman advance at the Medway, the Britons withdrew to the Thames and prepared to defend the crossing at London. Instead of launching his attack at once, Plautius sent word to Rome so that the Emperor could witness the decisive battle. Claudius hurried to join his legions, accompanied by a retinue of Roman aristocrats and a troop of elephants. Once he had arrived, the fighting could begin. It did not last long. Togodumnus was killed and his brother Caratacus fled to Wales. Within days, Claudius entered Colchester, capital of the Catuvellauni, surrounded by his elephants, to receive the submission of eleven British kings, including the Pictish King from Gurness in Orkney.

Claudius stayed in Britain for just over a fortnight, then returned to Rome, where he insisted the senate proclaim an official 'victory' and commission the building of a triumphal arch. From then on he insisted on being called 'Britannicus'. It did the trick. Claudius had gone from despised idiot to military hero in only six years. Even though the Emperor had returned to Rome, the invasion continued on and off for another forty years. As far as possible, actual fighting was restricted to tribes who did not submit voluntarily. In fact, this was easier than it seemed. The defeat by the Romans of the expansionist Catuvellauni was a cause for celebration among rival tribes and many of them viewed the Romans as liberators rather than conquerors. The Atrebates of Hampshire, the Iceni of Norfolk and the Brigantes of Yorkshire were happy to submit and pay their taxes rather than fight. After four years, Plautius had enlarged the frontier to the

Fosse Way. The only real resistance came in the Isle of Wight and Dorset, where the II Augusta, under the command of the future Emperor Vespasian, was forced to storm and capture some twenty hill forts from the Dumnonii before Vespasian could build his own legionary fortress at Exeter.

The first phase of the invasion was finished and, were it not for the perennial difficulty of establishing a stable frontier, it may have settled at that. To celebrate the orderly incorporation of Britannia into the Empire, an enormous monumental arch, 26 metres high, was built at Richborough where the Romans had landed. It was dressed in white Carrara marble and decorated with statues and inscriptions. Richborough stood on a promontory, so the arch must have been visible for miles out to sea. Its purpose was to emphasize that the Romans had tamed Britannia and every official visitor to the province entered through this arch before making his way inland along Watling Street.

However, further west things did not go so smoothly. The repeated attacks by the Silures, inspired by the fugitive Caratacus, persuaded the Romans that they must invade Wales. The first attempt was stalled when Suetonius Paulinus, who had routed the Druids on Anglesey, was forced to divert his troops to put down the far more serious revolt of the Iceni under Queen Boudicca. The Iceni had been a relatively quiet client kingdom under Boudicca's husband, Prasutagus. As a willing ally of Rome, it was his expectation that his kingdom would remain intact after his death. This did not happen. His property was seized, the aristocracy expelled from their estates and crippling taxes enforced. When she protested, Boudicca was flogged and her daughters raped. Her vengeance was swift and terrible. Rallying her own tribe, the Iceni, and the neighbouring Trinovantes in revolt, she swept through southern Britain, sacking and burning Colchester, London and St Albans. She tortured and killed every Roman and every Roman sympathizer that she could capture. The IXth legion, which tried to halt her advance, was cut to ribbons.

At the time of Boudicca's uprising, the south-east was considered to be well on the way to submission, so the bulk of the Roman army had been moved to the western front, from where Suetonius Paulinus was forced to abandon his invasion of Wales and return to deal with the revolt with the three remaining legions. If the uprising had been bloody, the retribution of Suetonius was even more so. Tacitus reckons 70,000 were killed on both sides during the revolt itself, and 80,000 during its suppression. Nero, who had succeeded Claudius as Emperor, seriously considered abandoning Britannia as a colony altogether.

After Boudicca's revolt had been put down, Roman control recovered. The crippling taxation was relaxed a little and those parts of Britain that had been conquered began the long process of assimilation into the Empire. But the stability of the northern frontier was beginning to crumble. Cartimandua, Queen of the Brigantes and the woman who had handed over the fugitive Caratacus, lost control of the loose federation of northern tribes. Agricola responded to this instability by pushing the frontier back to the very edge of the Scottish Highlands. He took his army even further north in his campaign against the Picts, inflicting a crushing defeat at the battle of Mons Graupius in AD 83. The location of Mons Graupius has eluded historians and archaeologists alike. The best guess is at Bennachie, near Inverurie on the banks of the River Don, 15 miles north-west of Aberdeen.

For the Romans it was a long way from home – 'the place where the world and nature end', according to Tacitus. But even with this defeat, the Highland Picts avoided being forced to submit to Rome in the way the Welsh did not, although the intention to complete the invasion of Scotland was there. At Inchtuthil, near Blairgowrie, a huge legionary fortress began to take shape, the equal of Chester or of Caerleon on the Welsh frontier. But reverses on the continent forced the Emperor Domitian to withdraw his troops from Scotland. The fortress was carefully dismantled and the materials taken south. It had been a lucky escape for the Picts.

By AD 120 the frontier had moved south to the line between the Solway Firth in the west and the mouth of the Tyne in the east. At first only a turf rampart, the frontier was turned into the impenetrable stone barrier of Hadrian's Wall, on the orders of the Emperor. Under Hadrian's successor, Antoninus Pius, the frontier moved north again. This time it was defined by the eponymous Antonine Wall, a barrier of rock and turf 20 feet high running between the Firths of Clyde and Forth. This was a much shorter boundary and many military historians think Hadrian should have built his wall here in the first place. But more trouble with the Picts convinced the next Emperor, Marcus Aurelius, to bring the frontier back down to Hadrian's Wall in 163. Even that great barrier was not impermeable and there were repeated raids across the wall as far as York.

Further south the fighting was less intense and the native population became drawn into the seductive and deliberate process of civilization. Towns were planned and built. Urban life, unknown in the whole history of the Isles, was born. People began to learn Latin and Roman dress became popular. As Tacitus shrewdly observed, 'Little by little there was a slide towards the allurements of degeneracy; assembly rooms, bathing establishments and smart dinner parties. In their inexperience the

Britons called it civilisation when it was really all part of their servitude'.

In the south, cities like Lincoln, Colchester and Gloucester grew up explicitly to accommodate army veterans on their retirement. Britons joined the army as auxiliaries and retired as citizens. In the towns, administrators mixed with craftsmen and artisans. Slaves were freed and were set up in business by their former masters. In the countryside, undefended villas of sumptuous magnificence sprang up, complete with wood- or coal-fired central heating, windows and glazed tile flooring. But even as these outward signs of affluence amused their owners, the seeds of destruction had been sown. The traumas of the Empire, its division into eastern and western sectors, the movement of the centre of the western Empire from Rome, first to Milan, then to Trier in eastern France, the deadly rivalries and murderous conspiracies all spelled the eventual end of the Roman occupation of Britain.

The surprise is that the Empire in the west lasted as long as it did. Even after extremely serious reverses – such as in AD 367 when a concerted assault by Picts, Saxons and Franks attacked the Roman provinces of Britain and Gaul, ranging at will, burning, killing and looting as they went – still the Romans managed to stage a comeback, this time under the Emperor Valentinian. By then the Roman army had changed its composition, no longer relying on Italian legionnaires or auxiliaries from the east. A quarter of the regular army was Germanic. By the beginning of the fifth century, the signs of weakening central direction were growing. There were no more bulk imports of coins, a sure sign that the army was not being paid as it once had been. The thriving pottery industry suddenly ceased. By AD 430 coins were no longer in regular use – another indicator of an ossifying economy. Even though there is evidence of one last attempt to reclaim Britain around 425, it came to nothing. By 450 Britain was well and truly on its own.

What lasting genetic legacy of the Roman occupation should we look out for? Whatever it is, we must expect it to be more pronounced in England, which was far more integrated into the Empire than Wales or Scotland ever were. And in Ireland we should not expect any significant traces at all. If Tacitus and other historians are to be believed, tens of thousands of Britons were slaughtered in the early years of the occupation – at least 80,000 as a result of Boudicca's revolt, 30,000 at Mons Graupius. These are large numbers for a relatively small population. The genetic legacy of wholesale military slaughter will be found, one imagines, mainly among men. The effect will be to reduce the diversity among Y-chromosomes. Population numbers can recover quickly if the women are spared, with men taking advantage of the surplus of women to bear their multiple children. But, with a smaller number of fathers, the Y-chromosomes that are passed down to future generations will not be as varied as if there were equal numbers of men and women.

The genetic origin of the Roman army itself is also something to be aware of when we examine the genetics. It was certainly not 100 per cent Roman, in the Italian sense, and drew its recruits from many different parts of the Empire, particularly from the lower Rhine. But nothing from the Roman occupation, save perhaps the import, and export, of female slaves, seems likely to have had a big impact on the maternal genealogy of England.

16

SAXONS, DANES, VIKINGS AND NORMANS

The end of the Roman occupation of Britain was quite unlike our own recent colonial goodbyes. There was no lowering of the flag, no salute from a member of the imperial family, no tear brushed away from the eyes of the last governor and no dignified departure on a warship. That was Hong Kong in 1997, not Britain in the fifth century AD. The Romans left a country already accustomed to the intermittent attention of raiding war parties from across the porous land borders to the west and north. In the great attack of 367, Picts from Scotland had joined Saxons from across the North Sea in rampaging through the countryside, killing and looting at will. The final withdrawal of the Roman army, some fifty years later, left England completely undefended and the population unprotected. Four centuries of occupation, during which citizens and slaves alike were forbidden even to carry arms and all weapons and military equipment were in the hands of the army, had left a population unaccustomed to warfare. That is not to say that the population was necessarily completely defenceless. Everyone must have seen this coming, and there were unknown numbers of retired veterans living in the towns and countryside. There may even have been remnants of a command structure at York and around Hadrian's Wall. The wall was not breached by the Picts, who must, therefore, have taken to the sea to attack the North Sea coasts in the great rising of 367. There were already Germanic settlements in eastern England based on former auxiliary units of the Roman army.

It takes only a little imagination to see these men using even their small advantages to establish themselves as minor kings in the confusion. But what actually happened is shrouded in mystery for one very good reason. There are simply no contemporary records. Even allowing for their exaggerations and creative imagination, the histories of Tacitus and others were some sort of record. After AD 410 there is nothing. We have to wait over 100 years for the next account – and that makes Tacitus sound as reliable as the *Encyclopaedia Britannica*. *The Ruin of Britain*, written by the monk Gildas in about 540, which we encountered in an earlier chapter, is little more than an indignant rant against the corruption and godlessness of his own time. This is how he describes the incursions of the early fifth century:

> As the Romans went back home, there eagerly emerged from the coracles that had carried them across the sea valleys the foul hordes of Scots and Picts, like dark throngs of worms who wriggle out of narrow fissures in the rock.

As to their appearance, Gildas writes, 'They were readier to cover their villainous faces with hair than their private parts with clothes'.

Here he describes the emergence of Vortigern from the chaos as a leader of the British and his invitation to the Saxon Hengist to protect him against Pictish attacks:

> To hold back the northern peoples, they introduced into the island the vile unspeakable Saxons, hated of God and man alike ... of their own free will, they invited in under the same roof the enemy they feared worse than death.

The Ruin of Britain is certainly colourful stuff – and totally unreliable. But what a great title. It is largely through the writings of Gildas that the central enigma of the Saxon age and its genetic effect on the British has been formed. Were they all killed or driven to the hills? This is what Gildas has to say about the effect of Saxon attacks in Norfolk:

> Swords flashed and flames crackled. Horrible was it to see the foundation stones and high walls thrown down ... mixing with holy altars and fragments of human bodies, and covered with a purple crust of clotted blood ... There was no burial save in the ruin of houses or in the bellies of the beasts and birds.

However, the archaeological evidence for immediate and wholesale destruction is conspicuously absent. London was not sacked, York and Lincoln were evacuated, then quickly recovered. In the far west the former legionary town of Wroxeter near Oswestry in Shropshire was completely untouched.

The far more dependable Bede, writing from the monastery at Jarrow, completed his *Ecclesiastical History of the English People* in 731. It is thanks to him that we are able to differentiate between the three tribes of 'barbarians', namely Saxons, Angles and Jutes.

According to Bede, Jutes from the Jutland peninsula of northern Denmark occupied Kent and the Isle of Wight, while Saxons from Saxony in north-west Germany settled in southern England. They eventually differentiated into the East Saxons, in Essex, the Mid-Saxons farther west (and remembered in the now vanished county of Middlesex) and the West Saxons of Wessex, which was much later divided into Hampshire, Wiltshire and Dorset. The Angles, originally located in Angeln in southern Denmark, between Saxony and Jutland, took over East Anglia, as well as the Midlands, which became Mercia, and Northumbria in the north-east.

In very broad terms, archaeology confirms Bede's account of the origins of the invaders, as far as the general area goes, with objects found in English graves of the period very similar to the styles of northern Germany and southern Denmark. But the neat division between Saxons, Angles and Jutes and their various destinations in England almost certainly applies only to the leaders, not the mass of settlers.

Unlike the 'barbarians' who finally defeated the Roman Empire within Europe, the Saxons, if I may use that term to embrace the three 'tribes' of Bede, came from well outside the frontiers of the Empire. They had completely different customs, and social organizations which emphasized kinship and loyalty to the chieftains. Honour was to be found in avenging the death of relatives, or accepting a payment, the *wergild*, in its place. The Gods were Norse – Tiw, Woden, Thor, Freya, and are remembered in the days of the week – Tuesday, Wednesday, Thursday, Friday – and also in English place-names like Tuesley in Surrey and Wednesbury in Staffordshire.

There was stiff resistance to the Saxons, culminating in the British victory around AD 500 at Mons Badonicus, an unknown location in the West Country where Geoffrey of Monmouth has King Arthur lead the victorious Britons. In the century that followed, the Saxons advanced only very slowly into territory still held by the Britons. By 600 the Saxons had moved north from Northumbria to defeat the Britons of southern Scotland. The Saxon victory at the battle of Chester in 616 severed the land link between the Britons of Wales and the Britons of the north, preventing them from helping each other. The British kingdoms of Rheged on the Solway Firth and Elmet around Leeds were extinguished, while Strathclyde, with its base in Dumbarton on the Clyde, survived. At the other end of the country, Cornwall resisted until the beginning of the ninth century. Saxon lands coalesced into larger kingdoms – East Anglia, Kent, Sussex, Essex, Middlesex, Wessex, Mercia, and Bernicia and Deria, both in Northumbria. Gradually, through conquest and alliance, kings of one region claimed sovereignty over one or more of the others. Raedwald of East Anglia, whose treasures were found at the burial site at Sutton Hoo, was one of these, claiming supremacy over Mercia and Northumbria.

Life in the court of Raedwald and other Saxon kings centred around the Great Hall and Bede gives a captivating account of what it was like: 'the fire is burning on the hearth in the middle of the hall and all inside is warm, while outside the wintry storms of rain and snow are raging'. The king, his earls and household listen to the songs and poems of their bards. This is the world of *Beowulf* – heroic, courageous and at the same time sensitive to literature and beauty, as even a brief glimpse at the Sutton Hoo treasure confirms.

One enduring question is why it was that the Britons did not simply absorb the invaders. This is what happened in France, where the Germanic invaders were quickly assimilated into the culture of Roman Gaul. Their language was almost entirely lost as Gaul slowly moved from Latin to French. But in England the reverse happened. English owes very little to Celtic, but almost everything to its Germanic roots. The abrupt change of language, the reason indeed that I am writing this book in English rather than a form of Welsh, is a major reason among historians and archaeologists for supporting the extermination scenario. Reading the bloodthirsty accounts from Gildas and faced with the extinction of the Celtic language and its replacement by English, it is tempting to explain them as variations on the theme of genocide. The English Celts were simply wiped out, or driven to the hills. Whether this is true or not is certainly something I hoped genetics would be able to discover, but is it really very likely?

There certainly were civilian massacres, on the eve of the battle of Chester in 616 for example, but there is also plenty of evidence that the British were living peacefully in Saxon kingdoms. A set of laws promulgated by a seventh-century king of Wessex specifically provides for Britons living in his territory. There is also the question of numbers. Is it realistic to think that there were enough invaders coming across the sea completely to supplant the native population? The genetics should provide a big clue towards resolving the perennial Saxon/Celt debate, and it is the main question to be answered about England. Or is it?

In the year 789 it is recorded that the King of Wessex married the daughter of the Mercian King Offa. Almost as an afterthought is added this ominous sequel:

And in his days came first three ships from Horthaland and then the reeve [the King's sheriff] rode thither and tried to compel them to go to the royal manor, for he did not know what they were; and then they slew him. These were the first ships of the Danes to come to England.

This was a chilling prelude to yet more raids, invasions and warfare by the mixed hordes of Vikings and Danes. After two centuries without any substantial foreign invasions in England, it looked as if it was starting all over again. After the killing of the king's sheriff in 789, on what has all the appearance of a reconnaissance mission, the Vikings paid most attention to the north of Britain and to Ireland, as we have already seen. But this was only a temporary respite. In 835 there was a large raid in Kent, then annually after that until, in 865, there was a full-scale invasion. The Danish Great Army landed in East Anglia led by Ivar Ragnusson, better known as Ivar the Boneless. I have rather a soft spot for Ivar the Boneless, because he was said to have suffered from the same genetic disease which I once researched myself. He was born, so it is said, with 'only gristle where his bones should have been'. From this description, Ivar almost certainly suffered from osteogenesis imperfecta, an inherited form of severe brittle-bone disease. If Ivar was anything like the osteogenesis patients I got to know he would have been very short, unable to walk without aid and with badly deformed limbs and spine. His head, however, would have been of normal size and his mental functions not impaired in the least.

The mystique of a fully mature mind in the broken body of a child is very powerful. I am not surprised that, even with this great physical disability, which would have prevented him from any combat himself, he was able to command an army by his legendary wisdom and force of personality alone. He was carried into battle on a shield. It must have been a disconcerting sight for the enemy.

Ivar forced the East Anglian king to supply him with food, horses and winter quarters, and next spring marched his troops north and captured the Northumbrian capital of York, beginning the long association between this city, renamed Jorvik by Ivar, and the Vikings. The Great Army then moved south to invade Mercia, then east to complete the invasion of East Anglia, which culminated in the brutal murder of Edmund, the Anglian king who had supplied the Great Army when it first landed. In three short years the Saxon kingdoms of Northumbria and East Anglia had been utterly destroyed.

The rampaging Great Army then turned south and prepared to invade Wessex. For the first time, the Danes were defeated, on the Berkshire Downs near Reading by Alfred and his brother Aethelred. The Danes withdrew and attacked again, this time beating the Saxon force near Basingstoke. The Danes were reinvigorated by the arrival of a new army in 871 and then prepared for the final showdown with the Saxons, with Alfred at their head. Alfred's Wessex and Mercia under King Burgred were the only Saxon kingdoms left in England that were not under Danish control. The Danes left Alfred alone for five years, and headed north, conquering Mercia *en route* to Yorkshire, which they began to divide up into permanent settlements. Then, at last, the Great Army turned south to attack the remnants of Saxon resistance in Wessex. They crushed Alfred at Chippenham in 878 and forced the king to retreat to his refuge in the marshes of Somerset, where he spent the winter arranging reinforcements. In the spring of 879 he headed towards Wiltshire and engaged the Danes at Edington Down on the slopes of Salisbury Plain near Warminster. He crushed the Great Army completely and forced their commander Guthrum to come to terms. The treaty separated England into two halves, with the dividing line running roughly north-west from London to the coast near Liverpool. East of the line was the Danelaw, to the west was Alfred's Saxon England. Schoolchildren learn that Alfred the Great saved England from the Danes. He clearly did not, as the Danes won control of half the country. Unsurprisingly, the peace did not last. Another army landed in 893, but restricted its campaign to the Danelaw and left Alfred's kingdom undisturbed.

From the genetic point of view I could see it was going to be hard to distinguish between Saxon and Dane. They both came from roughly the same place, their cultures were very similar, built around the Great Hall ideals of *Beowulf*. It was beginning to look, from the genetic point of view, like just another layer of north Germans and Scandinavians.

The next century saw the gradual reconquest of the Danelaw by the Saxon kings of Wessex. There were the inevitable setbacks. Norse armies recaptured York in 939 and 947, on the latter occasion under the command of the colourfully named Eric Bloodaxe. Another Danish army, under the equally chromatic Harold Bluetooth, had to be bought off after defeating an English militia in Essex. That only encouraged more raids, and by the turn of the first millennium huge amounts of cash had been paid to the Danes as what amounted to protection money.

The Vikings also used the same methods on the other side of the Channel. In 911 Hrolfe of Norway, or Rollo as he is more commonly known, sailed up the Seine and blockaded the river. In exchange for lifting the siege and withdrawing the threat to attack Paris, Rollo demanded, and got, a grant of land on the north-west coast from the French king. He became the first Duke of Normandy. He, his followers and descendants soon immersed themselves in French language and culture, though never forgetting their Viking roots.

Meanwhile, in England, the endless wars between Saxon and Dane continued. King Aethelred ordered a massacre of all Danes in England in 1002 – an impossible task, but serving to spread more hysteria and violence. Danes in Oxford took refuge in a church, but the citizens burned it down with the Danes still inside. The attempted ethnic cleansing forced Sweyn, the King of Denmark, to intervene, which he did on two unsuccessful campaigns until, in 1013, he launched a full-scale invasion. Aethelred fled to Normandy and thus began the fateful alliance that was to lead directly to the Norman Conquest. On Sweyn's death the following year his son Cnut, or Canute, inherited the Danish throne. By 1016 he had crushed Saxon resistance and become King of England as well. Notoriously he is the monarch who sat on the beach commanding the tide to retreat as a show of strength, but it was actually done to demonstrate his limitations in the face of nature. He was, in fact, a surprisingly good king, even though he divided his time between England and Denmark. But the fortunes of Wessex, whose regal supremacy Cnut had terminated, revived as Godwine, the Earl of Wessex, rose to prominence, even though he was not of the royal house.

Cnut died in 1035 and was succeeded by his son Harold. When Harold passed away five years later, his brother Harthacnut reigned for two brief years before he too died in 1042. That was the end of three decades of direct Danish rule and the kingdom was once more under a Saxon king, Edward (the Confessor), the son of Aethelred. Edward had grown up in Normandy at the court of his father-in-law, Richard, Duke of Normandy, after Aethelred had fled to France to escape the Danes in 1013. Already the Saxon royal family owed a debt to the Normans, a debt which only increased when the Earl of Wessex, Godwine, defied the king and threatened to seize control. To deflect this ambition of

Godwine, according to Norman propaganda, Edward, who had no children, promised the succession to William, Duke of Normandy. When Godwine died, he was succeeded to the earldom of Wessex by his son Harold who, again according to the Norman version of events, promised to back William of Normandy's claim to the English throne.

However, as he lay dying, Edward named Harold as his successor and England's last Saxon king came to occupy the throne on 5 January 1066. As William, Duke of Normandy, prepared the ground for invasion to press his claim to the Crown of England, the Danes were getting ready to do the same. Harald Hardraade, whose claim to the throne came through Cnut, was the first to attack. He invaded Northumbria and occupied York. King Harold, whose main army was in the south anticipating William's invasion from Normandy, was forced to move north to deal with Hardraade. This they did, and destroyed the Viking army at the battle of Stamford Bridge, close to York, on 25 September. Hardraade was killed. Three days later, on 28 September, William landed with his army at Pevensey Bay on the Sussex coast. Only nineteen days after defeating the Danes at York, Harold's exhausted army arrived to confront William at Senlac Hill, near Hastings. On the morning of 14 October 1066 the battle forces lined up. Harold's Saxon army massed behind a wall of shields on the crest of the hill and threw back charge after charge by William's heavy cavalry. In mid-afternoon, sections of Harold's army broke away to pursue a feigned Norman retreat and, without the advantage of the high ground, were cut off and overwhelmed. Harold was killed by an arrow and the day was lost. His men did not surrender, but fought to the death. They were all killed.

Having survived nearly three centuries of almost continuous

attack by Vikings from Norway and Denmark, the Saxon dynasty of Alfred eventually succumbed to the Vikings from France. The resistance had lasted from the day in the summer of 789 when the King's sheriff was murdered on the Dorset coast, to the death of the last of Harold's huscarls on an autumn afternoon 277 years later. In the 940 years since the Norman Conquest many have tried to invade the Isles, but none has succeeded.

17

THE DNA OF ENGLAND

Our strategy for recruiting volunteers for the Genetic Atlas Project in England was at first the same as our successful campaigns in Scotland, through blood-donor sessions. We did try one or two other methods, like setting up stalls at agricultural fairs. This worked very well in Cornwall, where a team at the nearby University of Plymouth was conducting a medical research project which involved taking blood samples. In other places it was less successful, mainly because we had not developed our cheek swab method for collecting DNA and were asking for blood. A further factor was that the majority of visitors to agricultural shows were farmers. On the one hand, this was why we had originally thought of the shows as a good place to collect DNA samples, reasoning that we were more likely to encounter families whose roots in the surrounding countryside went back a very long way. That part of the logic turned out to be true. But the flaw was that most farmers are men. If there is one universal truth which many years of fieldwork has taught me, it is that men are far more reluctant to give a DNA sample in unfamiliar surroundings than are women. In the blood-donor clinics it was different. We were part of the main event, not separate as I felt we were at the shows. If we had been a little more patient and perfected our approach, and if we had tried again with our cheek swab method, it may have worked. I think the reasoning was right; in Cornwall, where this worked really well, practically everybody whose DNA was sampled had at least two grand-parents from the local area.

Our first English blood-donor sessions were in East Anglia and over a couple of months we collected almost 1,500 blood samples in ways with which you are now familiar. I well remember travelling over the flat lands of the fens, where the soil is almost black and raised dykes drain the excess water to the sea. Scattered farmhouses and the occasional windbreak of Scots pine are all that protrude above the deadpan flatness. I know many people love the feel of openness and the big skies of East Anglia, but I need hills. When asking donors about their own origins, I was surprised how little movement there had been in the fens, even at places only a few miles from cosmopolitan cities like Cambridge. On one visit to a blood-donor session at the market town of Chatteris, 15 featureless miles north of Cambridge, the insularity of the fens came home to me when I was sitting next to a man in his forties, a farmworker, and going through my introduction of why we were there and why we wanted some of his blood. I explained how we were building up a genetic map and that was why we needed to know where people came from, so we could locate them properly on the map. When I had finished, he said he thought he probably shouldn't take part. That was unheard of, so I asked him why. He said it was because he had moved into Chatteris only recently.

'That's no problem,' I replied cheerfully. 'We can put you on the map wherever you're from, so long as we know. Where did you come from, before you came to live in Chatteris?'

'From Wimblington,' he replied.

'And where is Wimblington?' I enquired, ready to be told it was in Yorkshire or Dorset or somewhere else a long way away.

'It's up the road towards March,' he replied. And so it is, by 5 miles!

I think of that episode from time to time. The man is almost certainly still living in Chatteris and whenever I am asked how on earth I could ever expect to compile a genetic map from today's inhabitants that will reveal anything about the distant past, since people are so mobile these days, I tell them about the man from Chatteris.

Things were going nicely. We had more or less completed our DNA collections from Scotland, Wales and East Anglia and we had just been awarded another two years' funding from our major sponsors, the Wellcome Trust, which would give us ample time to complete our collections from the rest of England. I had arranged with other blood-transfusion regions in England to continue our work along the same lines. Word had got round that we did not interfere with the smooth running of the donor sessions. Indeed, donors on the whole enjoyed hearing about our work and it added a little more interest to their visit. I had a wonderful team who had honed their skills with, by now, three years of practice. In particular two of them, Emilce Vega and Eileen Hickey, who were assigned full time to the Genetic Atlas Project, were literally irresistible.

Nobody, male or female, young or old, could refuse Eileen and Emilce. They were, and still are, both striking young ladies, but in utterly different ways. Both are tall and slim, but while Eileen has the bright blue eyes, pale skin and auburn hair of her Irish ancestors, Emilce has the dark hair and deep brown eyes of her Argentinian forebears. Travelling to donor sessions with Eileen and Emilce was always interesting and our arrival at the small hotels we regularly used was always eagerly anticipated, and not because the owners were glad to see me again. Yes, things were going very well. Then disaster struck.

In scientific research the way is rarely smooth. Funds can be withdrawn, labs may have to be moved, extra duties of teaching or administration can be suddenly announced. It was none of these things. I put it all down to Ally McBeal. She, for those of you who do not know the TV series, was a glamorous New York lawyer, though prone to fits of hysteria and some very strange dreams. Suddenly a career in law became a very attractive option for young women. Two of my team announced that they were abandoning their scientific careers to retrain as lawyers. And one of them was Emilce. It's always sad to see that happen, but it is also very understandable. Despite all the publicity about how badly the country needs scientists, the prospects for young scientists are actually pretty dismal. Even if you succeed against very stiff competition in landing a junior academic position with the chance of a career in science, the pay is not good. With the upsurge in biotechnology in the late 1990s, law firms were keen to recruit and retrain geneticists for work in that sector as either patent or commercial lawyers. I could hardly object, and I did not. Soon afterwards, Eileen decided to move into forensics, which at least offered the prospect of long-term security, which young scientists crave. Of course, I cannot really blame Ally McBeal, but the loss of my two best fieldworkers was a blow. By the time I had recruited replacements for Eileen and Emilce, there were only ten months for the project to run. It was too late to get the new recruits up to speed on the delicate technique of charming the DNA out of blood donors.

So I decided to fall back on Plan B. This had its origins in an unexpectedly fruitful visit a few years previously to a service station on the M6 motorway, where I had first seen an advertisement for the electoral roll in electronic form. As this ad was in the toilets I was a little doubtful, but I ordered a copy anyway. It has been extremely useful and I have used it extensively in my genealogy work, tracing the names and addresses of men who share the same surname. Plan B aimed to recruit volunteers for the Genetic Atlas Project in parts of England we now no longer had the time to visit through blood-donor sessions. We could have written to people in the regions of England we needed to cover and asked for their help directly. But there were two predictable drawbacks here. First, this would be unsolicited mail with very little context and likely, as with other unexpected material, to dive head first into the wastepaper bin. The second problem was that we would have been unable to tell by their addresses alone whether people were new arrivals to an area or whether they had lived there all their lives. Although we could have placed the origins of new arrivals elsewhere in the Isles, which would still have been useful, we really needed people with deep roots in the area, like Chatteris Man, to fill in the large gaps that we still had left in our coverage of England.

We got round this by combing the electoral role for surnames. By choosing names which we could tell from their geographical distribution were local to the areas where we needed coverage, we stood a good chance of getting hold of volunteers who had, at least on their father's side, been there for several generations. Thanks to the strictly enforced feudal system instigated all over England after the Norman Conquest, estates had insisted that men adopt surnames. This was so that they could be told apart and so that inheritance of land tenancies from father to son could be properly controlled. By the end of the thirteenth century the practice had spread throughout the land, and practically everyone in England had a forename and a surname.

The logic of Plan B was that if we had a DNA sample from a man whose surname we knew was concentrated in an area we needed to cover, his Y-chromosome had probably been in the vicinity since the thirteenth century. For this to work, we needed a lot of names, for the following reason. Men with the same surname often have the same Y-chromosome signature, precisely because they are related to a common ancestor. It would be no use recruiting lots of men with the same surname just because they all lived in an area we needed to cover. Like as not, most of them would have the same Y-chromosome. To give you an extreme example, we could have got more than enough DNA samples from the Colne Valley in West Yorkshire just by writing to men with the surname Dyson. But 90 per cent of Dysons have the same Y-chromosome, owing to their common ancestry. We would get plenty of one particular Y-chromosome fingerprint and precious little else, and so our impression of Colne Valley genetics would be very misleading. For Plan B to give results for the Colne Valley that did give a representative picture of the whole area, we had to get DNA from all the local surnames. We would need to write to Dysons, Bamforths, Sykeses, Hirsts, Sutcliffes, Hills, Woods, etc., etc.

We worked our way through England, region by region, picking out scores of different surnames that, from directories and census distributions, were local to an area. We wrote to ten of each with an explanation of our project, a DNA sampling brush and return envelope. It was an exercise of, for us, military proportions.

We sent out over 15,000 DNA brushes and got just over 3,000 back, a return of a little over 20 per cent, which proved to be a remarkably consistent average whichever region we tried. We also sent out 5,000 brushes to addresses in Wales, with a similar 20 per cent return rate. The DNA from our earlier blood spots from Wales had proved difficult to extract for Y-chromosomes. Plan B was no substitute for collecting in person, but we were left with little alternative if we were to complete the project on time. We did eventually manage to fill in all the gaps in our coverage of England. Fortunately, the DNA brushes kept their precious cargo in good condition, even after several days in the post, and we had barely a failure when we set out to recover it in the lab. What did we find?

We were expecting, as you would too knowing its turbulent history, that England would be the most mixed of all the regions of the Isles. That is largely how it turned out, with the exception of Orkney and Shetland. In these Northern Isles, the settlement of so many Vikings had an enormous influence on what, from the Pictland results, we might imagine the genetic make-up of the indigenous islanders to have been. In the Northern Isles, the great surprise had been that the proportion of Norse women who settled was on a par with the men. That unexpected result came from the comparison of maternal and paternal lineages. Would we see the same sort of family-based settlement in England? Or would the genetics parallel the more lurid histories in seeing a massive replacement on the male side and very little on the female? Let's take a look.

I first divided England into the rural districts shown on the map (page xiii). The maternal clan pattern is stubbornly familiar wherever you are, but it does show a definite trend from the east and north to the south and west. It is literally as if the separation followed the line of the Danelaw. The Helena fraction is high, as usual, varying from 43 per cent in East Anglia to 47 per cent in the north of England. Below the Danelaw line it is only fractionally higher, rising to 49 per cent in the far south of England. There is really nothing in it. But it is in the other clans that the differences stand out, particularly when you get down to the detail, from which I will spare you. The most striking are the differences within the Jasmine clan and the presence of some very unusual sequences in East Anglia and the north of England.

Taking the Jasmines, the 'farming clan', to begin with, there are two different branches, which arrived in northern Europe by separate routes, as we saw in an earlier chapter. Let's call one the Ocean branch. They travelled around the coast of the Mediterranean from the Balkans, round Italy, to Iberia and then up the coast of France. The other, which we will call the Land branch, made their way overland to the Baltic and North Sea coasts. In Wales, Ireland and Scotland, the only branch is the Ocean branch. Only on the eastern side of Britain do I find much of the Land branch, and that is not a great deal. The great majority of Jasmines are from the Ocean branch and they pepper the map of the west side of Britain from bottom to top. They also occur in Norway.

The other difference in the matrilineal DNA is the occurrence of the minor clans of Wanda, Xenia and Ulrike. Wanda, along with Isha and Xenia, was originally subsumed in the clan of Xenia, and Ulrike is, as we saw, the 'eighth' daughter of Eve. All three are found in East Anglia and the north of England, but hardly anywhere else. Mrs Archer from Great Dunmow and Mrs Peachey from Coggeshall, both in Essex, are descendants of Ulrike. Ulrike's clan is particularly frequent in Scandinavia, so the hint is there that perhaps these two ladies are descended from that rare commodity, Viking women. Rare, that is, outside the Northern Isles. Xenia's clan originated in the steppes of Russia 25,000 years ago and travelled to Britain from the east. Wanda's clan is usually coupled with Xenia's, but has a more recent origin, 18,000 years ago, though she too came from the same vicinity. Mrs Lewis from Braintree in Essex and Mr Simmonds from Toft's Monk near Bury St Edmunds are both in Wanda's clan. They have certainly come a long way from the Ukraine.

In England there is a definite suggestion through detailed matches in most maternal clans of female immigration into the east from continental Europe, something which is undetectable in the west and north. How about the men? Here we do see a huge difference, even in the distribution of the clans, the crudest of indicators. Oisin's clan is down to only 51 per cent in East Anglia. The proportions increase as you travel west to Wales and north to Scotland. Where Oisin declines, Wodan increases and it reaches its highest proportions in the whole of the Isles in East Anglia, where Oisin is lowest. But there are virtually no Sigurds in East Anglia. However, there are plenty of Sigurds in the north of England, where they amount to 7 per cent of the total, which is a third of the Shetland total. In the south of England and in the Mercian territory of central England there are plenty of Wodans, and Sigurds too. The Appendix gives the figures.

The difference between the eastern regions and the rest intensifies when we look at the Y-chromosome diversity, which is much higher in the east, indicating a longer settlement if you follow the traditional way of interpreting genetic diversity. Diversity is much higher in the Wodan clan than in Oisin wherever you care to look.

By now there are so many threads in the air, so many facts to digest. And I have only been able to give you a tiny fraction of the detail. For every fact I have shown you, I have a hundred more in reserve. It has been a long tour, in time as well as in space. We have travelled to every corner of the Isles. At each step we have moved closer to an answer and now the time has arrived to distil the essence of our discoveries and draw our conclusions.

18

THE BLOOD OF THE ISLES

You have read the myths about the origins of the Isles that shimmer in the background, just out of reach; stories of brave kings and treacherous villains, fantastic monsters and invincible warriors. You have heard the ancient tales that have floated down the generations, stories that have been told and retold a thousand times around the campfire or in the flickering flame-light of the Great Hall. You have also heard how they were set down by Christian monks, transcribed from the world of the spoken and the sung to the realm of the written.

These same monks also wrote their own versions of our origins, histories that were sometimes an earnest attempt to pass on an impartial narrative of events and sometimes a fantastical torrent of loathing and contempt, fantasy and corruption. You have heard how these twisted histories were seized upon by kings, re-cast and put to work to bolster a fading reign or to right an ancient wrong and, in so doing, to inspire and justify a new conquest.

You have heard the chronicles of historians, from the amiable and conscientious Tacitus to the malignant architects of the Third Reich, each in their own way deriding and denigrating the people of the Isles as degenerate and barbaric. Yet archaeologists, whose account you have also heard, draw a sketch of the ancient Britons as masters of the shore and forest, able to fell the mighty aurochs with the well-directed flight of a flint-tipped arrow. You have heard how a medical man, the epitome of the Victorian amateur scientist, ranged through the Isles with card, tape and calipers searching for clues to our origins. You have heard how the search continued in the blood banks and laboratories of great hospitals.

I have introduced you to a new art and a new language. An art that is written in the codes of our DNA, those unseen architects of our bodies, even of our souls. It is a new art, not long tested and yet somehow irresistibly correct. How can anyone doubt that we are all our parents' children, as they also are the children of their parents? That is the simplicity of this art even though the language is new and obscure. I have tested you with talk of 'DNA sequences', 'haplotypes' and 'genetic diversity', of 'Y-chromosomes' and 'mitochondrial DNA'. I have impudently claimed that my art is oblivious to the prejudice of the human mind.

You have read the book, and I congratulate you on persevering through the technical sections. I have tried to make things as simple as I reasonably can, but it is no easy task to walk the tightrope between obsessive detail and arrogant patronage. My subject has been our history, the history written in our genes. Why, you might reasonably enquire, is this at all important in this day and age? What does it matter to me, you might say, whether my ancestor was a Viking, or a Saxon or a Celt? What difference will this make to my journey to work, what I eat for lunch or what I read on the way home? But if you really thought that, you would not have got this far. I hope you are by now just as fascinated as I am that within each and every one of our cells is something that has witnessed every life we have ever lived. I know that you can see the myriad threads of ancestry falling away beneath you into the abyss of the past.

I have introduced you to the brightest and strongest of these threads, one through which we are joined to our ancestral mother. An infinite umbilical cord which courses smoothly from mother to mother back into the mist of our ancestry. The other, which only men possess, thrusts its way from generation to generation. Erratic, illogical and passionate, it lives a life free from responsibility. But it enslaves its host and drives him to violence, murder and conquest. Follow this thread into the past at your peril. Sooner or later you will spend a generation or two in the testis of a warlord. We could not have any more different conduits into the depths of our ancestry.

The stories that these threads tell are completely individual. They are not composites or averages. I have been at pains to point out, even to the point of repetition, that to squeeze them through the mangle of mathematics risks robbing them of their vitality, silencing their murmurs. What I have tried to do is to listen to the whispered stories of thousand upon thousand of these threads and to divine patterns from the swirls. Enough of the philosophy – what are these patterns?

The first conclusion, blindingly obvious now I can see it, is that we have in front of us two completely different histories. The maternal and paternal origins of the Isles are different. And that should be no surprise, given the opposing characters of the chroniclers. The matrilineal history of the Isles is both ancient and continuous. I see no reason at all from the results why many of our maternal lineages should not go right back through the millennia to the very first Palaeolithic and Mesolithic settlers who reached our islands around 10,000 years ago. The average settlement dates of 8,000 years ago fit with that. But that cannot be the complete answer. That was well before the arrival of farming, and the presence, particularly in Ireland and the Western Isles, of large numbers of Jasmine's Oceanic clan, and her companions from the maritime branch of Tara, says to me that there was a very large-scale movement along the Atlantic seaboard north from Iberia, beginning as far back as the early Neolithic and perhaps even before that. The number of exact and close matches between the maternal clans of western and northern Iberia and the western half of the Isles is very impressive, much more so than the much poorer matches with continental Europe.

That is not to say this was a 'wave' arriving all at once and swamping the small numbers of Mesolithic inhabitants of Mount Sandel, Starr Carr and the like. They were well established, knew the land inside out and must have been easily able to adapt, gradually, to a less mobile agricultural existence. The change from hunter-gathering to agriculture may have taken centuries or millennia. There is no archaeological evidence of conflict and no reason to suppose that the arrival of the farmers would have been confrontational, at least not at first. We encountered the peaceful co-existence of Mesolithic and Neolithic communities in Portugal where the new arrivals from the Middle East cleared the woods for their crops well away from the coastal zones favoured by the residents. I think this pattern would have been reproduced all over the Isles. There was plenty of room, with the Mesolithic population only a few thousand strong and with plenty of land available for cultivation after the woods had been cleared. The mere presence of large numbers of Oceanic Jasmines indicates that this was most definitely a family-based settlement rather than the sort of male-led invasions of later millennia. I think the main body of the Neolithics arrived by this western route, since the Oceanic Jasmines reached right round the top of Scotland to the east coast and even inland to the Grampian region. There are far fewer Land Jasmines in the Isles. I found none in Ireland, only one in Wales, just five in Scotland, again in the Grampian region and in Strathclyde. The rest are in England and concentrated there in the Midlands and the east.

After that, the genetic bedrock on the maternal side was in place. By about 6,000 years ago, the pattern was set for the rest of the history of the Isles and very little has disturbed it since. Once here, the matrilineal DNA mutated and diversified, each region developing slightly different local versions, but without losing its ancient structure. Without agonizing over the precise definition, this is our Celtic/Pictish stock and, except in two places, it has remained undiluted to this day. On our maternal side, almost all of us are Celts.

I can see no evidence at all of a large-scale immigration from central Europe to Ireland and the west of the Isles generally, such as has been used to explain the presence there of the main body of 'Gaels' or 'Celts'. The 'Celts' of Ireland and the Western Isles are not, as far as I can see from the genetic evidence, related to the Celts who spread south and east to Italy, Greece and Turkey from the heartlands of Hallstadt and La Tène in the shadows of the Alps during the first millennium BC. The people of the Isles who now feel themselves to be Celts have far deeper roots in the Isles than that and, as far as I can see, their ancestors have been here for several thousand years. The Irish myths of the Milesians were right in one respect. The genetic evidence shows that a large proportion of Irish Celts, on both the male and female side, did arrive from Iberia at or about the same time as farming reached the Isles. They joined the Mesolithics who were already here, having reached the Isles either by the same maritime route or overland from Europe before the Isles were cut off by the rising sea.

The connection to Spain is also there in the myth of Brutus, who came to the Isles from the Mediterranean and up the Atlantic coast to found New Troy in the land of Albion. This too may be the faint echo of the same origin myth as the Milesian Irish and the connection to Iberia is almost as strong in the British regions as it is in Ireland.

One myth that the genetic evidence certainly does not support is the relic status of the Picts. Their ancestors, just like the rest of the people of the Isles, have been there a very long time, but they are from the same basic stock. They are from the same mixture of Iberian and European Mesolithic ancestry that forms the Pictish/Celtic substructure of the Isles. It is very clear from the genetic evidence that there is no fundamental genetic difference between Pict and Celt.

This ancient matrilineal bedrock has been overlain to any substantial extent in only two places. In Orkney and Shetland there was a large settlement of women from Norway during the Viking period and the ancestors of roughly 40 per cent of today's Shetlanders and 30 per cent of modern Orcadians first stepped ashore from a Viking ship. But plenty of others in the Northern Isles can trace their ancestry back well before the Viking age to the sophisticated Picts who built the brochs at Mousa in Shetland and Gurness in Orkney.

The second overlay is in eastern and northern England, above the Danelaw line which ran from London to Chester. Above that line, and particularly in the east, there are clear signals of female settlement overlying the Celtic substratum. As we have already touched on, it is very difficult to distinguish Saxon, Dane and Norman on a genetic basis, since they are all from the same Germanic/Scandinavian origins, but the concentration of these signals above rather than below the Danelaw line makes me think they are more likely to be Viking than Saxon or Norman. The approximate extent of this overlay I estimate to be between 10 per cent in the east and 5 per cent in the north – substantial in terms of numbers, but really only denting the Celtic substructure.

Lastly, I have found a tiny number of very unusual clans in the southern part of England. Two of these are from sub-Saharan Africa, three from Syria or Jordan. These exotic sequences are found only in England, with one exception, and among people with no knowledge of, or family connections with, those distant parts of the world. I think they might be the descendants of Roman slaves, whose lines have kept going through unbroken generations of women. If this was the genetic legacy of the Romans, they have left only the slightest traces on the female side. I have not found any in Wales, nor in Ireland and only one in Scotland. This is an African sequence from Stornoway in the Western Isles, for which I have absolutely no explanation. These exotic dustings, and the more substantial layers of Viking maternal lines, are the exception. Everything else in the Isles, on the maternal side, is both Celtic and ancient. But what about the men?

Here again, the strongest signal is a Celtic one, in the form of the clan of Oisin, which dominates the scene all over the Isles. The predominance in every part of the the Isles of the Atlantis chromosome (the most frequent in the Oisin clan), with its strong affinities to Iberia, along with other matches and the evidence from the maternal side convinces me that it is from this direction that we must look for the origin of Oisin and the great majority of our Y-chromosomes. The sea routes of the Atlantic fringe conveyed both men and women to the Isles. I can find no evidence at all of a large-scale arrival from the heartland of the Celts of central Europe among the paternal genetic ancestry of the Isles, just as there is none on the maternal side.

The pockets of ancient Wodans in mid-Wales and the 'Pictland' regions of Grampian and Tayside are, I believe, the echoes of the very first Mesolithic settlers who arrived from continental Europe, perhaps even travelling by foot while there was still a land connection. They look old to me, and for an apparently contradictory reason. That is because they are all very similar. The same applies to the Oisins. And yet the customs of genetics state that the longer a gene has been in a place, the more diversity should have accumulated. That was how I was able to fix the homelands of the seven European clan matriarchs. Using that rule I placed them at the locations where the present-day diversity was highest, and thus where they had had longest to accumulate mutations away from the original.

But this rule does not seem to work with the paternal lines delineated by the Y-chromosome. The very striking thing about the clan of Oisin throughout the Isles is how very similar they all are. Or at least, how there are very large clusters of very similar chromosomes in one location, and not in others. For instance, the Ui Neill chromosome reaches a very high frequency in north-west Ireland but is rare elsewhere, and the Somerled chromosome is common in the Highlands and the Hebrides, but virtually unknown elsewhere – unless carried by a member of Clan Donald or Clan Dugall. This dramatically reduces the genetic diversity, and leads to very recent settlement dates, sometimes obviously incorrect. This has been noticed before with the Y-chromosome but has been attributed to what is called 'patrilocality'. This is the practice of men staying put, while the women move to marry. However, I don't think this works well enough to explain the amazing similarity in the Oisin chromosomes. The explanation is less cosy.

This is the 'Genghis effect' and it is not confined to the Mongol Empire. In the Isles very large numbers of men, perhaps all of them in the clan of Oisin, are descended from only a few genetically successful ancestors. All the conditions are here in the Isles. From the Iron Age onwards, and certainly during the first millennium AD, which we have covered here in such detail, the past is filled with the continual feuding between rival clans. One of the genetic consequences of the rise of powerful men is that they monopolize the women and have more children. I have even argued in *Adam's Curse* that therein lies the motivation for their procreative ambition. We can see the evidence in the Isles in the Scottish clans of Macdonald and Macleod and in the Irish Ui Neill. These are very dramatic examples of a process which has percolated throughout the history of the Isles. That is why the diversity has been lost. It is because only comparatively few men have left patrilineal descendants. So, the longer a clan has been in a place like the Isles, the more similar the Y-chromosomes become. That is the reason our Celtic Y-chromosomes are so alike.

It is also the reason why most of the Wodan chromosomes are the opposite. They are usually very diverse indeed in the Isles. Not because they have been here a long time, but because they are comparatively recent. There are pockets of 'old' Wodans in Wales and Pictland, but in the east and in the north above the Danelaw line, the Wodans, which reach 31 per cent in East Anglia, are extremely varied. I scarcely found any two the same when I looked at the detailed fingerprints, unless they had the same surname and were thus related to a common ancestor through that route. The clan of Oisin still predominates in every part of England, but the bedrock is substantially overlaid in the east. Because of the genetic similarity of Saxon, Dane and Norman, I cannot discriminate so easily between them. But I estimate that approximately 10 per cent of men now living in the south of England are the patrilineal descendants of Saxons or Danes, while above the Danelaw line the proportion increases to 15 per cent overall, reaching 20 per cent in East Anglia. Only a few of these men have surnames of Norman origin and, taking this into account, I estimate the Norman Y-chromosome legacy at 2 per cent or below even in the south of England.

From this evidence the succession of Saxon/Danish invasions during the turbulent centuries after the Romans departed did leave a mark on the stubbornly Celtic indigenous bedrock of parts of England. It is a real presence, but it is by no means completely overwhelming. The gory chronicles of Gildas do contain a grain of truth. The roughly twofold excess of Saxon/Danish Y-chromosomes compared to their maternal counterparts hints at a partially male-driven settlement with some elimination or displacement of the indigenous males. But the slaughter, if slaughter there was, was not total and still there are far more people with Celtic ancestry in England, even in the far east, than can claim to be of Saxon or Danish descent.

I have tried to find Roman Y-chromosomes, but they left very few traces that I can be sure were theirs. Only one very rare patrilineal clan, without even a name, may be the faint echo of the first legions. It is found in southern Europe, including Italy. What makes me think, as well as this link to Italy, that it might be linked to the Romans is that it is entirely restricted to England. There are no traces beyond the borders with Wales or Scotland.

There may be others, but as was pointed out, the tradition of recruiting legionaries and auxiliaries from Gaul and other parts of the Empire, as well as from Britannia itself, makes them very difficult to spot among the descendants of later arrivals from the same areas. But true Roman genes are very rare in the Isles.

Overall, the genetic structure of the Isles is stubbornly Celtic, if by that we mean descent from people who were here before the Romans and who spoke a Celtic language. We are an ancient people, and though the Isles have been the target of invasion and opposed settlement from abroad ever since Julius Caesar first stepped on to the shingle shores of Kent, these have barely scratched the topsoil of our deep-rooted ancestry. However we may feel about ourselves and about each other, we are genetically rooted in a Celtic past. The Irish, the Welsh and the Scots know this, but the English sometimes think otherwise. But, just a little way beneath the surface, the strands of ancestry weave us all together as the children of a common past.

This genetic history has been read mainly from the surviving genes passed on by generations of ancestors who lived through the events described in *Blood of the Isles* and whose descendants carry them today. It is a new history, reconstructed from thousands of fragments from the past. The general conclusions in this and other chapters have been distilled from the DNA of hundreds of people from each region of the Isles. But to the people concerned, and to everyone else in the Isles, it is our own genetic ancestry that is the most important. It is the thread that goes back to our own deep roots that means the most. The proportions of one clan or another are vital and the detailed genetic comparisons are essential for arriving at any sort of general conclusion, but it is our own ancestry that understandably, and quite rightly, holds the most interest. Now that we know what the overall patterns mean and now that we can identify with confidence surviving DNA with the different ancestral signatures, it is open to anyone to find their place in this amazing story. For this is not the history told by fading manuscripts in dimly lit libraries, or by rusting weapons in glass cases. It is a living history, told by the real survivors of the times: the DNA that still lives within our bodies. This really is the history of the people, by the people.

APPENDIX

Within this appendix I have compressed just a small fraction of the genetic data from the Oxford Genetic Atlas Project that forms the foundation for *Blood of the Isles*. More details appear at www.bloodoftheisles.net.

Distribution of Maternal Clans in Scotland (%)

	Argyll	Borders	Northern Isles	Tayside	Grampian	Highland	Hebrides	Strathclyde	All Scotland
Helena	51.2	41.7	53.1	47.4	46.7	38.9	31.6	44.9	45.3
Isha	4.9	2.8	2.6	4.6	5.0	6.1	7.0	1.4	4.2
Jasmine	6.5	19.4	10.8	15.3	20.6	13.1	14.0	14.0	13.4
Katrine	9.8	5.6	5.1	6.1	6.1	4.5	14.4	7.2	7.2
Tara	9.8	2.8	7.7	10.2	6.1	10.6	13.6	15.5	10.1
Uma	0.8	0.0	0.0	0.0	2.2	1.0	0.0	1.0	0.6
Uta	0.8	2.8	0.0	0.5	1.1	15	3.5	1.9	1.2
Ulrike	3.3	8.3	4.4	2.6	2.8	2.0	0.9	1.4	2.8
Ursula	6.5	11.1	10.6	5.1	5.0	10.6	9.2	7.2	8.4
Velda	1.6	2.8	2.2	4.1	2.8	6.1	2.6	4.3	3.3
Wanda	0.8	0.0	0.4	1.0	0.0	1.5	0.4	0.0	0.6
Xenia	2.4	2.8	3.1	2.6	1.1	2.6	2.2	1.0	2.3
Other	1.6	0.0	0.0	0.5	0.6	1.5	0.4	0.0	0.5

Distribution of Maternal Clans in England and Wales (%)

	North-umbria	North	Central	East Anglia	London	South-west	South	All England	North Wales	Mid Wales	South Wales	All Wales
Helena	51.0	47.1	45.8	43.0	40.9	45.8	48.6	45.7	52.1	42.7	47.2	46.3
Isha	3.0	4.7	3.8	4.6	3.7	2.6	3.4	3.9	6.3	7.3	0.0	6.1
Jasmine	15.0	13.7	11.0	9.1	13.4	13.7	12.3	12.2	11.5	8.5	2.8	8.8
Katrine	12.0	9.0	8.0	7.6	6.1	7.9	8.4	8.2	6.3	11.0	11.1	9.5
Tara	2.0	7.9	10.2	8.8	11.6	7.4	12.3	8.9	4.2	11.0	19.4	9.8
Uma	0.0	0.5	0.8	1.8	0.0	1.1	0.0	0.8	1.0	0.6	2.8	1.0
Uta	0.0	0.5	0.4	0.3	0.0	1.1	0.0	0.4	0.0	2.4	2.8	1.7
Ulrike	1.0	1.1	1.5	5.5	1.2	3.7	0.6	2.3	5.2	0.6	0.0	2.0
Ursula	10.0	8.8	13.6	10.7	14.0	11.1	8.9	10.9	10.4	11.0	11.1	10.8
Velda	5.0	3.6	2.7	4.3	3.0	2.1	1.7	3.2	3.1	3.7	0.0	3.0
Wanda	0.0	3.0	1.5	1.8	3.7	0.5	0.0	1.8	0.0	0.6	0.0	0.3
Xenia	1.0	0.0	0.8	2.4	0.0	1.1	0.0	0.8	0.0	0.6	0.0	0.3
Other	0.0	0.0	0.0	0.0	2.4	2.1	3.9	0.9	0.0	0.0	2.8	0.3

* The clans of Isha, Wanda and Xenia are minor clans subsumed within Xenia. Uma, Uta and Ulrike are minor clans in western Europe, but more frequent further east.

Distribution of Paternal Clans in Scotland (%)

	Argyll	Borders	Northern Isles	Tayside	Grampian	Highland	Hebrides	Strathclyde	All Scotland
Oisin	81.1	78.1	59.9	78.9	83.5	75.9	71.2	73.3	72.9
Wodan	3.8	12.5	16.8	17.5	11.8	16.5	17.8	20.0	15.4
Sigurd	7.5	3.1	19.8	1.8	2.4	6.3	11.0	4.2	8.8
Eshu	1.9	1.6	1.5	1.8	2.4	1.3	0.0	1.7	1.5
Re	5.7	4.7	1.5	0.0	0.0	0.0	0.0	0.0	1.2
Other	0.0	0.0	0.5	0.0	0.0	0.0	0.0	0.8	0.3

Distribution of Paternal Clans in England and Wales (%)

	North-umbria	North	Central	East Anglia	London	South	South-west	All England	North Wales	Mid Wales	South Wales	All Wales
Oisin	68.3	62.8	65.8	51.2	57.6	57.7	78.2	64.0	78.5	86.4	84.2	83.2
Wodan	15.9	25.1	21.4	31.2	23.2	36.4	12.6	22.2	15.0	8.2	10.5	11.0
Sigurd	7.3	7.2	7.1	2.4	4.0	2.5	4.2	5.2	2.8	0.7	0.0	1.4
Eshu	1.2	1.9	0.0	3.2	3.3	4.9	1.3	2.1	3.7	2.7	2.6	3.1
Re	2.4	2.5	1.5	5.6	4.0	3.1	1.7	2.7	0.0	1.4	0.0	0.7
Other	4.9	0.6	4.1	6.4	7.9	5.5	2.1	3.6	0.0	0.7	2.6	0.7

CLAN DISTRIBUTION – MATERNAL

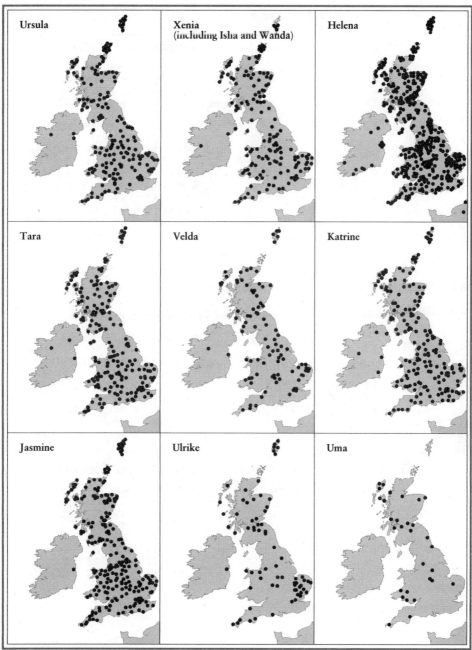

Locations shown are the birthplaces of the paternal grandfathers of Oxford Genetic Atlas Project volunteers.
No independent Irish data shown.

CLAN DISTRIBUTION – PATERNAL

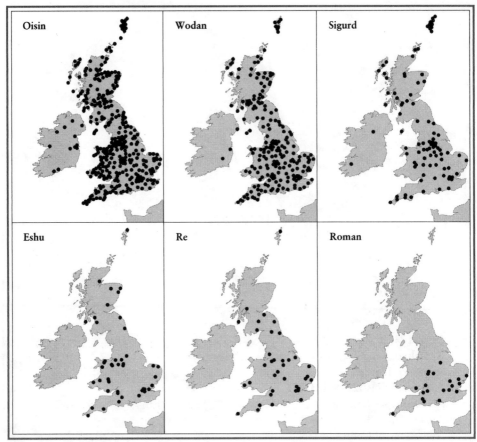

Locations shown are the birthplaces of the paternal grandfathers of Oxford Genetic Atlas Project volunteers.
No independent Irish data shown.

N.B. These data are copyright. If you make use of them please acknowledge the source as
Sykes, B.C., *Blood of the Isles*, Bantam Press, London (2006).

INDEX